EX LIBRIS

POSSESSING
GENIUS

POSSESSING GENIUS

THE BIZARRE ODYSSEY OF EINSTEIN'S BRAIN

CAROLYN ABRAHAM

PENGUIN

VIKING

VIKING

Published by the Penguin Group

Penguin Books Canada Ltd, 10 Alcorn Avenue, Toronto, Ontario, Canada M4V 3B2

Penguin Books Ltd, 80 Strand, London WC2R 0RL, England

Penguin Putnam Inc., 375 Hudson Street, New York, New York 10014, U.S.A.

Penguin Books Australia Ltd, Ringwood, Victoria, Australia

Penguin Books (NZ) Ltd, cnr Rosedale and Airborne Roads, Albany, Auckland 1310,
    New Zealand

Penguin Books Ltd, Registered Offices: Harmondsworth, Middlesex, England

First published 2001

10 9 8 7 6 5 4 3 2 1

Printed and bound in Canada on acid free paper ⊗

NATIONAL LIBRARY OF CANADA CATALOGUING IN PUBLICATION DATA

Abraham, Carolyn
    Possessing genius : the bizarre odyssey of Einstein's brain

Includes index.
ISBN 0-670-89221-1
1. Harvey, Thomas Stoltz 2. Einstein, Albert, 1879–1955. 3. Neurosciences. 4. Pathologists—
United States—Biography. I. Title.

RB17.H365A27 2001   616.07'092   C2001-901739-1

Visit Penguin Canada's website at **www.penguin.ca**

For my parents,
Dudley and Thelma Abraham,
who nurtured brain *and* heart,
and for Stephen Rouse, who takes
such good care of both.

The cult of individuals is always, in my view, unjustified. To be sure, nature distributes her gifts unevenly among her children. But there are plenty of the well endowed, thank God, and I am firmly convinced that most of them live quiet, unobtrusive lives. It strikes me as unfair and even in bad taste to select a few for boundless admiration, attributing super-human powers of mind and character to them.

This has been my fate, and the contrast between the popular estimate of my powers and achievements compared to the reality is simply grotesque. The awareness of this strange state of affairs would be unbearable but for one pleasing consolation: It is a welcome symptom in an age which is commonly denounced as materialistic that it makes heroes of men whose goals lie wholly in the intellectual and moral sphere.

This proves that knowledge and justice are ranked above wealth and power by a large section of the human race.

—ALBERT EINSTEIN, JULY 1921, AFTER
VISITING THE UNITED STATES OF AMERICA
FOR THE FIRST TIME

CONTENTS

	Introduction	*xi*
ONE	GENESIS	1
TWO	NOBLE AMBITION	22
THREE	'55-33	47
FOUR	PROMISES, PROMISES	69
FIVE	PIECES OF GENIUS	96
SIX	RESIGNATION	116
SEVEN	THE DOMINO EFFECT	125
EIGHT	LOST AND FOUND	140
NINE	FOUR TRIPS TO CALIFORNIA	163
TEN	INHERITANCE	193
ELEVEN	A DEEP, DARK SECRET	214
TWELVE	WORKING-CLASS HERO	231
THIRTEEN	METAMAN	258
FOURTEEN	CANADIAN CARTOGRAPHER	281
FIFTEEN	THE BIG BANG	308
	Epilogue	*343*
	Sources	*348*
	Acknowledgements	*371*
	Index	*375*

INTRODUCTION

I CAME LATE to the parade. Reporters had marched through the life of Thomas Harvey in intermittent bursts for forty-five years before I ever spoke to the man who took Albert Einstein's brain. Harvey was by then a full-fledged legend. Around him swirled half-truths and larger-than-life stories that only added to the mystique of the three cerebral pounds that reshaped our view of the universe. The basic tale is simple enough. Harvey was the chief pathologist at Princeton Hospital where Einstein died in 1955. The Nobel laureate's corpse ended up in the morgue and Harvey extracted the brain, hoping it would offer clues to solving the mystery of genius. But beyond that there is nothing simple about the story. A few years after dissecting the intellectual icon of the modern age, Harvey disappeared, and with him went the brain. It popped up through the decades in peculiar places—in a box under a beer cooler and in a bedroom closet, stashed in small-town Missouri and in the airtight obscurity of Tupperware.

On a winter night in 1996, Einstein's brain rumbled over the Canadian border, jiggling in two jars of alcohol in the trunk of a Dodge. I heard about the Niagara crossing three years later. In June 1999, an assistant to Hamilton neuropsychologist Sandra Witelson phoned the Toronto office of the *Globe and Mail*, where I write about medical research. She was calling to give us a heads-up. There was big news coming out of their lab at McMaster University: Witelson was about to publish the first anatomical study of Einstein's brain—and she had discovered in it something extraordinary.

The findings naturally fascinated me. Scientists have searched for physical evidence of intelligence since leeching was a common practice. They have inspected neural molecules and unwound reams of genetic code to find something tangible to explain what makes one person more brilliant than others. So why, nearly half a century after the death of the amazing physicist, was the first report on the structure of his brain only now being published? Anatomy studies hardly demand the wizardry of late-twentieth-century technology. They could weigh and measure just as well in the post-war era as they could on the cusp of a new millennium. I wondered what on earth had taken so long.

Old news clippings laid out the brain's shady history. I read about the scandal that stuck to Harvey's taking of the organ. I read about his suspicious departure from Princeton Hospital and the urban myths that sprang up afterward: some imagined him to be a mad hunchback scientist on the run; others a fugitive who had abandoned his professional ethics to pinch a relic. The datelines were always different —New Jersey, Kansas, Missouri . . . Harvey sounded like a gypsy traipsing America's backwaters with a pickled crystal ball, a thief who had helped himself to the contents of Einstein's skull.

Harvey always denied the accusations. He said Einstein's family had given him permission to study it. Trouble was, no statement from Einstein's family was ever found to verify his version of events. Anyone with direct knowledge to set the record straight had long since died, and so the suspicions never did. Was Harvey a thief? And if he was, how did the squadron of handlers, which springs up around any celebrity, fail so spectacularly to protect the physical remains of the biggest star science has ever seen? Armed Israeli soldiers stood watch over Einstein's papers as they were transferred from the United States to Jerusalem. The trustees of Einstein's estate were so fiercely protective of his reputation that it took more than thirty years to publish the first volume of the physicist's collected papers. Even the place where Einstein's ashes were scattered was for years a well-guarded secret, to keep worshippers from swarming the site. Yet the most precious part of the great thinker—his brain, which Einstein once called his "laboratory"—somehow ended up in the possession of one obscure man, free, apparently, to dole it out and display it to whomever he deemed fit. And if Harvey was a thief, why was he never caught? Did he take the brain for science, or as a souvenir? If it was an object of scientific study, why had so little science been done?

In the two years since Sandra Witelson published her report, I've been looking for answers. Some of them are surprising. Ever since Harvey slipped away with the genius brain, people have assumed he was a lone wolf, independent in his tenure as curator. But private letters, filed away at the Albert Einstein Archives in Jerusalem and never before published, tell a different, and somewhat tragic story. Harvey, it turns out, was not alone. Someone was always keeping tabs on the custodian of Einstein's brain; someone, not so different from Harvey, who was anxiously waiting to hear what scientists would have to say about it. But the wait turned out to be much longer than expected.

In many ways, developments in neuroscience over the last fifty years decided the fate of Einstein's brain. Harvey had meanwhile promised that anything discovered about the organ would be published only in scientific journals. Yet the power of Einstein's celebrity would make the promise impossible to keep. It utterly engulfed the rest of Harvey's life, but he could no more give the brain up than serious scientists could keep from swooning over its post-mortem pieces.

Most people who hear about the strange tale of Einstein's brain are torn between disbelief and laughter, as I was in those first moments contemplating the jars bouncing over the border. I hope the following pages offer another perspective, and that some questions, unlike the brain of Albert Einstein, can be laid to rest.

Carolyn Abraham
Toronto, June 2001

POSSESSING
GENIUS

ONE

GENESIS

☼

LATE ONE NIGHT in New York City, in an old brick building in the Bronx, Harry Zimmerman lay in a hospital bed waiting for sleep. All his life he had worked around hospitals. Now he was a patient, admitted to a private floor, tucked in a private room, surrounded by the sound of his own breathing. A gastric tube snaked through a hole in his neck, sucking fluids from his blocked intestines. Cancer of the colon, the doctors had said. He could picture the disease: polyps swelling like balloons in the lining of his bowels.

How many times had he held malignant tissue in his own hands? For twenty-four years he had run the pathology department at this very institution. He had joined in 1947, when the Montefiore Medical Center had specialized in treating patients with neurological disorders. After they died he would study their brains. Now, in 1995, he was dying. The chances of curing colon cancer plummet with age, and at ninety-three Zimmerman was nearly as old as the century itself, with hair now winter white and a salamander complexion. Comfort, that's the best a human can hope for at the end. He ripped the gastric tube from his throat.

A medical intern working the overnight shift appeared in the doorway.

"Let's put that back, shall we, Dr. Zimmerman?" he said.

Most of the staff knew him because he had often come to work at the pathology lab long after he'd retired. Sometimes, if he bumped into a medical student, he would deliver a spontaneous lecture on the anatomy of the brain. The son of a Ukrainian stonemason, he had been a director of Yeshiva University's medical school, a professor at Yale and Columbia, a man who trained a generation of neuropathologists. To colleagues he seemed immortal as he peered up from his microscope and the arcane world where human behaviour intersects with the cold architecture of nerve tissue, a fire in his eyes. Lucid to the last, that's what they would say about old Harry Zimmerman when he was gone. Obediently, he tilted his head back for the hands of the young doctor.

"It's three o'clock in the morning, Dr. Zimmerman, try to get some sleep," the intern said.

But Zimmerman didn't feel like sleeping. He felt like talking. So the intern sat down. "What's your name?"

"Salvatori," the intern said, "Roberto Salvatori."

In the quiet before dawn, while the rest of the city dreamed, there were hundreds of things Zimmerman could have told the young stranger. He might have said, as the old tend to do, that when he was Salvatori's age he had studied in Germany, at the epicentre of early research into the human brain. He might have regaled him with war stories, about sharing a tent with the future co-inventor of the polio vaccine, or made Salvatori laugh by explaining how he had prevented the spread of parasites among soldiers by prescribing clean sheets. Or he could have described the tour of duty in Guam during which he discovered that the island's Chamorro natives suffered one of the highest rates of amyotrophic lateral sclerosis, Lou Gehrig's disease, in the world. Post-mortems showed that some of their brains had shrivelled to three-quarters of their former mass. Some sections were as hollow as jack-o'-lanterns. Zimmerman became a leading expert on the condition. Maybe he assumed the young doctor already knew.

If he had turned sentimental and spoken of his long life with Miriam, it would have been understandable given the circumstances. They had never had any children; always, it had been just the two of them. But it's funny what a man thinks of when time is running out, and Zimmerman told the intern none of this. Instead, his voice grew solemn and intense—as though he were about to share a secret, Salvatori thought. Then Zimmerman began speaking of a man who had died forty years earlier, and the story of a brain that had weighed on his conscience ever since.

✹

It grew in the belly of a German woman in the summer of 1878. It was barely a speck beneath skirts and petticoats and flesh, drawing forest-scented air from the Swabian Alps northwest of the Danube. The woman lived there with her husband in a medieval city of narrow, winding streets called Ulm. The husband was a cheerful, meaty man of thirty-one who tended a fledgling electrical business near their apartment. She, eleven years his junior, brunette and broad of forehead, kept house and played piano when her chores were done.

Probably it was late August before Pauline Einstein realized it was growing. By then it had morphed into a neural plate, unfurled from a hollow ball of embryonic cells. A groove had run down the centre of the plate within a few days of its existence. Then the ends of it stretched slowly downward and curled toward themselves until they touched to form an enclosed tube of protoplasm—a jelly like no other in the world. Cells within the tube replicated. Some served the role of nursemaids, catering to those that would grow into mighty neurons. Together, not long after conception, they began to construct the magical scaffolding of human thought.

No one alive knows how Pauline Einstein conducted herself in its crucial early weeks. Perhaps, because this was her first child after two years of marriage, she sated her passions and cravings. Perhaps the brain blossomed while she banged out Beethoven and devoured the local delicacies of sausage and onion cakes and sweet apple wine. Hermann Einstein no doubt wished good things for his unborn child, but in all likelihood he did not pray for them. He and his wife did not attend the local synagogue or deny themselves pork. Such customs seemed ancient superstition to him. Their families were Swabian Jews of southern Germany who had passed down a laissez-faire attitude toward the less practical facets of life, religion included. So while masons erected the tallest church spire in Europe opposite his workshop, Hermann did not contemplate the will of God.

By the fourth week, 125,000 cells assembled along the groove in the middle of the neural tube, each an ancestor of the some 100 billion neurons that would eventually stun the world. They divided and multiplied, 50,000 per second. Then they drifted forth, like pioneers into a fluid wilderness, to settle the cerebral frontier. Primitive nerve cells clung to their nursemaids, the tiny, elongated glial cells whose radial feelers forged their paths. They literally wrapped tendrils around the nerve cells and inched like snails over and between those that had already reached their destinations. Chemical instructions in the form of hormones and proteins zinged back and forth, telling each nerve cell precisely where to let go and stake its claim. For the oldest ones, the journey was short, completed in a single day. But with every successive wave of migration from the centre, the voyage lengthened. Younger generations of cells travelled more than a week to reach the outer edges of the bulging fetal brain.

☼

There was something otherworldly about being in the presence of the dishevelled physicist whose daydreams redesigned the universe. People who met Albert Einstein never forgot it. They would vividly recall details about the day, the mood and even the weather. So it was with Harry Zimmerman, as he told the intern from his sickbed, about a caustic cold and dreary day in January 1953. Zimmerman had been fifty-two then, his career on a happy trajectory. He had left his teaching position at Yale after the war, suspecting that he had been passed over for promotion because he was Jewish. But Yeshiva University in New York had offered him the chief pathology job at the Montefiore Medical Center, then asked him a few years later to direct its new medical school. Zimmerman hoped it could be named after the most famous scientist in modern history. So he and four other Yeshiva officials drove to Princeton, New Jersey, to the small clapboard house where Albert Einstein lived. Zimmerman knew it would not be easy to convince the professor to lend his name to the project. Einstein, he told the intern, was a very modest man.

The more the world adored the genius, the more celebrity bewildered him. When thousands of Japanese fans held a night-long vigil below his Tokyo hotel room in 1922 and roared their approval when he appeared on the balcony the next morning, Einstein told his wife, "No living being deserves this sort of reception . . . I'm afraid we're swindlers. We'll end in prison yet." Humble protests that he was simply a curious man only deepened the veneration of strangers, most of whom understood nothing of the theoretical physics that had made him famous in the first place. They wrote to him from all corners of the globe, for advice on their marriages, for handwriting samples, for the meaning of life. Einstein often wrote back, to prison inmates and children and beleaguered wives. But he was stingy with his name and suspicious of those who sought to use it for their own purposes. Companies he had never heard of courted him to be their spokesman after he emigrated to America in 1933. They offered him fortunes to endorse their ties and toilet waters, their disinfectants and musical instruments, products he had never used. "Isn't it a sad commentary on commercialism?" Einstein said. "And I must add that business firms make these offers with no thought of wanting to insult me. It evidently means that this form of corruption is widespread."

Despite all this, Zimmerman thought the plan for a Jewish-sponsored medical school might appeal to the professor. A former student of Zimmerman's from Yale, a young doctor by the name of Thomas Bucky, had agreed to introduce him and the other Yeshiva officials to the genius physicist. Bucky's father, Gustav, was an American inventor and doctor who had known Einstein back in Berlin. The Bucky family had spent holidays with the scientist at his German summer home in Caputh, before he'd fled Hitler.

Einstein greeted them all cordially when they arrived, and Zimmerman politely waited out the small talk before broaching the reason for their visit. Predictably, Einstein seemed astonished when he did. He said he was no physician, that a medical school should be named for someone trained in medicine. Zimmerman pointed out that medicine and mathematics often converged. Still, Einstein was unimpressed. So Zimmerman tried another tack and told him that a refusal would force him to select Sigmund Freud as the namesake.

"But this will imply that your school is committed to psychoanalysis," Einstein said.

"Well, then we could name it the Sir William Osler School of Medicine," Zimmerman replied.

"But he is not Jewish."

"In this case we shall call it the Joseph Goldberger School."

Einstein looked troubled. "Who is Joseph Goldberger?"

"You know, Professor," Zimmerman said triumphantly, "no one would ever ask me, 'Who is Dr. Einstein?' "

A smile played around the edges of Einstein's mouth then, his eyes twinkled, and he asked whether Zimmerman had said he was a doctor or a lawyer.

Three months later, they met in Princeton again when more than a hundred people gathered in driving rain to celebrate the physicist's seventy-fourth birthday and the official naming of the Albert Einstein College of Medicine. On Einstein's seventy-fifth birthday, Zimmerman visited him at his house and showed him an architectural model of his namesake college. Through all this they became friends, Zimmerman told the intern, so that eventually Zimmerman felt comfortable enough to ask Einstein about his brain.

✷

It looked like a lima bean by the fall of 1878, just over a centimetre long. The upper curve jutted toward the face, rounding in the lower portion to form the remarkable eyeballs that would float one day in a New Jersey safety deposit box. Two hemispheres budded on either side of the central groove, and the whole organ swelled with activity. Nerve cells hurried to connect with one another, scaled each other, built the brain from the inside out. By the twenty-fourth week, the vast sweep of protoplasm changed from barren horizon to crowded forest as seventy thousand cells huddled in brain tissue the size of a pinhead and the final migrants arrived to craft the perimeter. Nerve cells pushed and shoved each other into columns six layers deep in some areas to form the cerebral cortex. As they clumped and shifted, the lima bean transformed into a walnut, its outer shell wrinkling and crumpling to allow its considerable surface area to fit neatly within the confines of the skull. Somewhere amid the valleys and ridges of the grey crust lay the potential to envision and reason, to calculate and write, to translate imagined observations of space and time into profound mathematical equations.

Nestled in their rightful place, the primitive nerve cells burst into neurons. Branches of dendrites sprouted across their cell bodies in the tens of thousands. They fanned out and feathered, like fireworks exploding, into the shape of chandeliers, spindles, baskets, pyramids and bouquets; as varied as the foliage in the woods beyond the uterine home. The dendrites awaited electrochemical signals from other neurons to fire up or calm down. Meanwhile, an axon grew like a trunk on each neuron to send the messages, to make synaptic connections with a dendrite, no matter how distant. They snaked through layers and lobes and criss-crossed the fibrous canyon of the corpus callosum that connected the brain's two halves. They stretched through the inner sea-horse form of the hippocampus that straddled the brain stem and several centimetres down, to the very base of the spinal column. The circuits exploded in a rat-tat-tat racket of electrical activity as cells in one quadrant fired off messages to cells in another, squirting chemicals between them, frantically forging trillions of connections to process the world. But half of the 200 billion neurons were surplus, an overabundance to guide the growing fetal body. Partway through prenatal life, tens of thousands of them committed mass suicide. They literally blew out their own cell walls, pulverized their own genetic material. Had all of them lived, inflating the brain to unseemly proportions, they surely would have retarded its function.

At the brain's core, structures common to every creature with a backbone kept the body alive. The fist-like cerebellum, the almond-shaped amygdala, the thalamus above the hypothalamus began shouting chemical orders to sleep and breathe, to pump blood and churn out hormones. Through the pons, a bridge of beautifully spiralling nerve cells, down through the medulla oblongata and into the spine, the brain reached out to the body. It monitored the heart, lungs, liver, stomach and kidneys and evaluated their fitness to survive outside the womb. On the cusp of spring, a March day in 1879, a chemical whisper from a pea-sized hypothalamus shot cortisol into the bloodstream and Pauline Einstein's contractions began.

☼

Zimmerman had learned under German neuroscientists how to slice brain tissue thin as prosciutto, dye it different colours and correlate what he saw to human disease. In Guam, he'd found dendrites and axons tangled in useless knots, which explained why the stricken Chamorros trudged with the stoop of old men, their muscles slack and their speech slow, their hands trembling like flamenco guitarists. He told Salvatori that he'd wondered whether Einstein's brain might offer a physical clue to intelligence, some "anatomical marker of exceptionality." Einstein's name had, after all, become synonymous with genius itself. His bolts of brilliance had toppled three hundred years of scientific thought. Even Nobel laureates puzzled at his ideas.

From 1905 to 1916, revolutionary notions about the secrets of energy, the substance of matter, the speed of light, the plasticity of time and space and the very composition of the universe spilled from his pen, and Einstein compared it to child's play: "A normal adult never stops to think about problems of space and time. These are things which he has thought of as a child. But my intellectual development was retarded, as a result of which I began to wonder about space and time only when I had already grown up." His visions seemed bold and fantastic. Then came evidence that they might also be right. In 1919, the skies offered proof for his theory of relativity, which posited that the pull of gravity was the result of curved space—not a force comparable to magnetism, in the way Newton had described it. Astronomers photographing a solar eclipse from Brazil and West Africa captured the light of distant stars bending, deflected as it passed the indentation of space around the sun. The measurements confirmed Einstein's view that space is like a three-dimensional cushion sagging beneath bowling balls of heavenly bodies. Anything that rolls near it falls toward it. The verdict was so unexpected that the *New York Times* had sent its golf reporter to cover the announcement.

Einstein's reputation in scientific circles had grown slowly over the previous two decades. There had been no crush of job offers, no crowds of reporters clamouring for interviews. Yet suddenly the post-eclipse headlines catapulted him to almost instant, international fame—an amazing phenomenon in an age before television. The press lauded his triumph, said he had overthrown Isaac Newton and launched a revolution. The *New York Times* wrote that "this news is distinctly shocking and apprehensions for the safety of confidence even in the multiplication table will arise." That so few could actually fathom Einstein's theories only added to his mystique. As he himself once put it, "I am sure that it is the mystery of non-understanding that appeals to them." People just knew he had something significant to say about the elements that shape reality, and it sounded vaguely romantic—talk of sun and light from distant stars, of supple space and the trickery of time, words more familiar to poetry or prayer than science. Einstein's friend and biographer Abraham Pais would write that in the wake of war, strikes, famines and Bolshevik uprisings, the obscure physicist emerged to expose a new order in the universe and became the quintessential celebrity, "famous for being well known."

Two and a half decades later, Einstein would be inextricably linked with the Allies' victory in the Second World War. "Fat Man" and "Little Boy" exploding over Hiroshima and Nagasaki thundered confirmation of his equation that mass multiplied by the speed of light squared indeed equalled energy, making Einstein the apologetic father of the atomic age. The once out-of-work teacher, relieved to have a job reviewing technical inventions at a Swiss patent office, had become a scientific hero: a galactic clairvoyant, grounded to earth and lesser mortals by his shabby sweaters and sockless feet.

Other scientists could not resist the notion that the brain beneath his unruly bouffant might somehow be a freak of nature. In 1951, researchers at Massachusetts General Hospital plastered electrodes across his scalp to record the electrical symphony of his brain cells speaking to one another, hopeful that they might discern a chord of creativity. With their trusty new electroencephalogram, they measured its idling current. The needle zigzagged impressive peaks while Einstein apparently figured quadratic equations in his head. Then suddenly the squiggly line flattened. Researchers rushed in to find out what the great physicist had been thinking to deaden the machine. Einstein apparently told them he had heard rain and remembered that he'd left his rubbers at home.

Zimmerman told Salvatori that the professor did not bristle at his request to examine his brain after death. Even if Einstein thought it futile to search for the biology of intelligence, he did not deny a scientist the right to look. Was it any more outlandish than having the temerity to challenge Newtonian thought and Euclidean space, to rewrite the laws of the universe? As Zimmerman recalled it, Einstein said, "My brain ought to be studied, so that if anyone has a question, it can be answered." His only condition was that there should be no publicity around such a post-mortem examination or any of its findings. Zimmerman told the intern that he had intended to respect Einstein's wishes and vowed not to publish the results of any study performed on the organ. But Zimmerman had not known then that the promise would never be his to keep.

✻

From the moment it feasted on its first gulps of oxygen, logging the sound and smell and taste of "mother," Einstein's brain became something different from what it had been in the womb. Neurons forged new connections to register each new sensation, blurring forever which features the organ had been born with and which ones a lifetime stamped upon it. His parents thought his head looked physically unusual from the start. Its size and angles horrified them. "Much too fat, much too fat!" his maternal grandmother wailed. They wondered if he was deformed, but the doctors told them time would make it right.

In his infant cortex, glial cells spun fatty white sheaths of myelin around axons to insulate them like electrical wires, creating a sea of white matter beneath the brain's grey coat. The sheaths speed impulses ten to one hundred times faster than they can travel on a naked axon, and those allowing him to see, hear, touch and move probably myelinated first, smoothing the baby's jerky motions.

His parents worried about his brain even after he had learned to walk, because Albert did not jabber new words as other toddlers did. He often repeated himself under his breath after he spoke, so that later scientists would wonder whether neural networks devoted to language had already ceded cortical ground to some other cognitive function. Einstein told biographers that he'd simply put off talking until he could speak in complete sentences.

The childhood brain was soft as custard. It would have spilled over and between the fingers of anyone who had tried to touch it then. In its nascent exuberance, it forged millions more connections between neurons than it could ever use. Each waited to be stimulated, directed to a certain task; to grip a pencil, plunk a piano, laugh. Neurons competed with each other for thinking jobs, and the losers perished in another ruthless round of pruning. Experience determined which ones survived, sculpting the brain into a unique pattern of emotion and thought. Nurture did not compete with nature; it waltzed with it as the boy wandered alone at the age of four through the hectic metropolis of Munich, dodging horse-drawn carriages and towering crowds. At the age of five, his brain puzzled at the compass his father gave him, transfixed by the invisible force that kept its needle pointed north. That same year, he tossed a chair at his fiddle teacher, but with practice the malleable young cortex obliged, so that within a decade he played Mozart on the violin.

Some schoolteachers thought Einstein dull-witted and moody, said he would never amount to anything because he responded to them with open disrespect for their authority. Yet Einstein had no difficulty concentrating when he chose to. His younger sister, Maja, always remembered how he could build a house of cards fourteen storeys high. The poor medical student his parents hosted for lunch on Thursdays would fascinate him too, introducing the young Einstein to books about science, astronomy and philosophy. His Uncle Jakob presented algebra to him as a "merry science in which we go hunting for a little animal whose name we don't know," and in these matters Einstein flourished, his young brain soaking up the language of mathematics, bending neurons to its logic. At twelve, he fell in love with Euclid, calling the text his uncle gave him his "holy geometry book." At thirteen, he began to study higher mathematics on his own, surpassing the knowledge of his own instructors. In the midst of family parties, he squeezed onto the sofa, balanced an inkpot on the armrest and worked feverishly on equations, oblivious to everyone around him. At sixteen, Einstein wrote his uncle a letter proposing an experiment to see if electricity, magnetism and the ether were connected. Already, synapses fired toward an extraordinary future.

✺

Zimmerman never told Salvatori how pieces of Einstein's brain came to be in his possession. The subject remained a sore point with him: the expectation, the excitement and then the disappointment. He felt thwarted—he had not been there on that spring day when they opened up the genius. Instead, it had been a younger pathologist, a student of his from Yale. So Zimmerman never held it in his hands, never saw it whole. By the time his sample of the brain arrived, in a wooden box fastened shut by a tin clasp, it had already been shaved into wafers. He had no control over what became of it through the years, no power to object if it huddled in a tea canister outside Tokyo, in a fridge in Honolulu, in a secret spot in Hamilton and in so many other places besides.

Zimmerman had written to a colleague in 1975 that he had studied the microscopic sections in search of a "genius cell" but had found
nothing of the kind. The only memorable fact was that the tissue
showed few signs of deterioration, despite seventy-six years of continuous labour beneath a cranium. Otherwise, he told Salvatori, the
brain looked "absolutely normal" to him. He dismissed the investigations of other scientists as though they could no more detect genius
in Einstein's post-mortem brain than they could find speed in the legs
of a dead runner.

☼

In its twenty-sixth year, the brain's performance peaked. An adoles-
cence spent digesting Copernicus, Kepler, Galileo, Newton's laws
of gravity and Maxwell's theories of electromagnetism had likely
beaten well-worn paths of cellular connections. Pulses of electricity
that can travel more than 300 kilometres per hour between neurons,
releasing dopamine, serotonin and acetylcholine chemicals to spark
other neurons, can take just one-thousandth of a second to recharge
and fire again. With the daring of youth, and whole pints of blood
coursing under his skull by the second, the brain manipulated mental
images at unknown speeds.

Only later, after the Nobel Prize had been won and fame cemented,
did the myths grow up around the ease of his thinking, bolstered by
the caricatures of a fuzzy-haired stargazer, puffing calmly on a pipe
while his fecund mind deciphered nature's great secrets. But it was
labour. In one rare admission, Einstein revealed that his urge to understand the universe kept him in a state of "psychic tension . . . visited
by all sorts of nervous conflicts. . . . I used to go away for weeks in a
state of confusion, as one who at that time had yet to overcome the
stage of stupefaction in his first encounter with such questions." But
he was so compelled to solve the questions that he could hardly keep

himself from his abstract thoughts. They distracted him from the grief of losing his father, tore him away from his new wife and first-born son. Even at the patent office in Berne, where Einstein once described himself as a "venerable federal ink shitter," he stole moments from the blueprints he reviewed to pull out pen and paper, the only tools of his laboratory, and set to work.

For two hundred years, scientists had assumed that light moved in waves, like ripples across a pond. But experiments around the turn of the century showing that light was capable of knocking electrons off metal sheets, literally scattering them from the surface, bruised this notion. The German physicist Max Planck, the father of quantum theory, suggested that tiny particles, or quanta, absorb and emit light so that it is actually matter that displaces the electrons. But Einstein pushed the concept even further. He claimed that light itself is made up of particles, discrete chunks of energy (later known as photons) that can batter the surface of metal and send the electrons flying. At the time, Einstein considered his hypothesis unprovable, yet this so-called photoelectric effect would win him his first and only Nobel Prize fifteen years later and lay the groundwork for the twentieth century's electronic achievements—radios and televisions, remote controls and motion detectors, machines to scan everything from carry-on luggage to groceries at the supermarket.

Einstein often found himself drawn to the problems that stumped other scientists. Later generations of physicists would marvel at his ability to distill complex ideas and inconsistencies to their simplest forms and then see them clearly. One such riddle was the mystery of Brownian motion, named for the British botanist Robert Brown, who, in 1827, noticed that grains of pollen suspended in liquid never sat still or settled to the bottom. The ceaseless jiggling enticed Einstein to explain this perpetual motion, which seemed to defy the laws of nature. Visualizing how the tiny grains drifted to and fro in a glass of water, he concluded that the particles were not dancing of their own accord; they were bombarded and ricocheting off other smaller particles, atoms, invisible even beneath a microscope. He mathematically illustrated how these finite molecules within the liquid actually pushed the pollen from place to place. His thought experiment not only helped to establish atomic theory but it presented scientists with a formula by which to chart the behaviour of atoms themselves. Einstein said he could actually envision his theories and equations in three-dimensional forms, that his thinking seemed a muscular, visual process. "When I examine myself and my methods of thought I come to the conclusion that the gift of fantasy has meant more to me than my talent for absorbing positive knowledge."

Of all his breathtaking cerebral feats, none would distinguish Einstein's brain more than the exploration of a thought he had played with since puberty. He used to wonder what the universe would look like if he could zip through space riding a beam of light. When he considered the question again in adulthood, he redefined Newton's laws of space, time and motion. Where the Newtonian world featured space as a flat field of ether on which all the elements of the universe play, the Einsteinian world featured space as a tricky container that curves around matter and exists only when there is something in it. Without a relationship to time or matter, space is merely the presence of nothing, immeasurable and undetectable. To the young man's thinking, space and time could not be separated, since neither were absolute measurements but changed with the vantage point of any observer, like the passenger parked at the train station who is convinced he is moving when another train speeds past. Only when the passenger sees a third, stationary object—a bag, say, on a platform—can he be certain that he is still. To the people queued on that same platform, a seven o'clock train arrives when the little hand of their watches reaches the number seven, but for people in another part of the world, the arrival of the train coincides with another hour altogether, so that the motion itself is completely relative to a person's position in space. Einstein felt that space and time depend wholly on one's frame of reference, shattering any idea that the notion of a "now" even exists. Light, he concluded, is the only stable element of the universe, travelling at the constant speed of 300,000 kilometres per second, regardless of the space and point of time it passes through, regardless of the vantage point of the observer. Did the train's beacon not shine on the waiting platform passengers at the same speed at which it illuminated the tracks ahead?

Sitting on the streetcar one day, Einstein imagined himself, as he had in his youth, straddling a light beam whizzing through space. Only now, he was staring back at a clock tower as he blazed away, and a wild realization seized him. The greater the distance he travelled away from it, the slower the clock ticked, until it stopped, frozen in time. Neural fireworks must have exploded at the thought. The light beam, he now understood, travelled at its constant speed while matter became smaller and heavier the farther he got from it, and the farther he got from it, the more slowly time passed. Time, he concluded, was simply a man-made invention, and a moving clock slows down, loses time—though it would be decades before space and air travel proved him right. For the befuddled public, Einstein diluted his special theory of relativity with trademark wit: "A minute sitting on a hot stove seems like an hour, but an hour sitting with a pretty girl passes like a minute."

His brain crackled with implications as his hands scribbled equations to the electrochemical square dance of his neurons. Like a seamstress he stitched his ideas together, proving that energy itself was merely matter and that all matter contained energy, leading him to formulate the most famous equation of all time: $E=mc^2$. He went on to picture himself in a falling elevator, and from that deduced the equivalence principle, which says that gravity and acceleration exert the same force, reinterpreting Newton's falling apple with the notion of bent space. By drawing his assumptions together, he explained the origins and architecture of the universe.

His work founded the study of cosmology. Scientists in the field wondered if they need ever conduct experiments again to prove their theories. Perhaps observable fact was redundant. Einstein seemed to demonstrate that the mind is the only weapon necessary to hunt the truth. Lost in his musings, scientists of a future time and space would wonder: had blood flooded his parietal lobes, at the top rear quarter of each brain hemisphere, as he worked? Was it from there, between neurons devoted to vision and body sensation, above the lower bulge of mathematical and spatial reasoning, that his genius sprang? Was the wondrous mind that unified forces of the cosmos itself merely matter?

☼

It had been Zimmerman's understanding that "at the express request of the family of Professor Einstein no written report [was to be] published of the findings on both gross and microscopic examination" of Einstein's brain. Zimmerman told the intern that was why he never wrote a thing about it, not that he had much to say. And as far as he could tell, that younger pathologist had kept the pact of silence as well for nearly three decades. Then, out of nowhere it seemed, interest swelled in neurology circles. A California team felt they had discovered a clue to Einstein's brilliance, a fellow in Alabama found something else, and God knew who else was looking. Scientists seemed to be vying for it like fans stretching to catch a home run ball. The media, meanwhile, had never lost interest in the study of that remarkable brain, and Zimmerman himself had slammed the phone down on reporters who called asking questions. In 1985, after Einstein's brain popped up on the west coast, a journalist from *Discover* magazine wanted to know why there had been a tug-of-war over the organ. "This is nonsense," Zimmerman fumed. "This is confidential information. I was told many years ago by the family to have nothing do with publicity."

Since then the brain had turned up again and again, always, it seemed, in a different location. Zimmerman probably couldn't help but think of the last official discussion he'd had with the executor of Einstein's estate. How troubled that old man had been about the uncertain fate of his friend's vaunted brain. So that now, in his own final days, it bothered Zimmerman's conscience, too. Would Einstein's wishes be fulfilled if the brain drifted among strangers through another century? Would science have lost anything if the brain had been incinerated and scattered into a New Jersey wind with the rest of him?

Sleep beckoned and Zimmerman fell quiet. There was nothing more to say. The intern closed the door softly behind him.

Colon cancer killed Harry Zimmerman on July 28, 1995, and Roberto Salvatori would remember their discussion long afterward.

☼

The male brain can shrink with age, gobbling less glucose than it did in its youth. Levels of dopamine chemicals that bounce among cells inciting pleasure can decline, limiting ability to think abstractly or solve problems. The biological facts haunt mathematicians, who fear their work will not improve beyond the age of thirty. In his book *A Mathematician's Apology,* G. H. Hardy wrote that "mathematics, more than any other art or science, is a young man's game."

At twenty-seven, after publishing an astonishing four landmark physics papers in a single year, his *annus mirabilis,* Einstein himself wondered whether he was washed up. Later he would say, "Truly novel ideas emerge only in one's youth. Later on, one becomes more experienced, famous and foolish." It is true that his brain disappointed him in the second half of its natural life. He searched obsessively for a theory to unify the laws of gravity and electromagnetism, a unified field theory, but found none. He carried a notepad to jot down thoughts whenever they struck him, while smoking a pipe on his front stoop or perched in his sailboat waiting for the wind to pick up. At the wedding of his friend Thomas Bucky, he penned furiously on a dinner napkin. Even in the hospital, between morphine shots to soothe the burning of the main artery leaking in his belly, he contemplated equations. Nurses found a page of them on the nightstand beside the bed where he died.

Einstein had devoted himself to theoretical physics at the expense of almost everything else, making him an absentee father to two sons, a distracted husband to two wives. In his later years, though, the brain expended a good deal of its cerebral energy cultivating a profound regard for the rest of humanity. The Holocaust, the war, the threat of nuclear weapons all ignited political longing for a one-government world and crusades for peace and social justice. Einstein's essays on these topics filled books, and his observations were quoted with the reverence shown to proverbs. The fatheaded boy, slow to talk, had evolved into a philosopher capable of sheer poetry. "Our situation on this earth seems strange," Einstein told the German League of Human Rights in 1932. "Every one of us appears here involuntarily and uninvited for a short stay, without knowing the whys and the wherefore. In our daily lives we only feel that man is here for the sake of others . . . . Although I am a typical loner in daily life, my consciousness of belonging to the invisible community of those who strive for truth, beauty, and justice has preserved me from feeling isolated. The most beautiful and deepest experience a man can have is the sense of the mysterious. It is the underlying principle of religion as well as all serious endeavour in art and science. He who never had this experience seems to me, if not dead, then at least blind."

Seventy-six years after his birth, the blood supply dried around the brain. It no longer gushed but surged and then finally stopped, like a car before it runs out of gas. The organ constituted about 2 percent of his body weight and consumed roughly a third of his body's blood.

Without a steady supply, the organ began to fail—its life-giving chemicals evaporating. The hippocampus, on the underside, parallel to his ears, where he might have stored old memories of a compass and directions to his summer cottage at Caputh, likely died first, as it is the most sensitive to lack of oxygen. Then, slowly, cortical neurons, starved of sugar, began to quiet. Brain cells can function for up to twenty hours after a heart stops beating. Meek signals might even pass among them, though no one knows what or if a brain thinks in its final moments. But within a dozen hours of Einstein's death, a metal wedge pried open its cranial case. A slippery membrane was peeled back from the brain's glistening surface and the hands of the first person who would touch it reached in to pull it out.

TWO

NOBLE AMBITION

❄

SECRETS COULD BE hard to keep in a small town like Princeton. The children found out the very next morning. On April 19, 1955, Mrs. Schafer asked her fifth-graders at Valley Road School if they had anything to contribute for current events. A smart girl sitting near the front shot her hand into the air.

"Einstein died," Katrina Mason blurted out, proud to be the bearer of such momentous news. Yet she had no sooner spit the words from her lips when the voice of an otherwise quiet boy carried from the back of the room: "My dad's got his brain."

If heads swivelled in disbelief or fell back in laughter at Arthur Harvey's revelation, Katrina Mason was too distracted to notice. She was mortified: to have imagined that reporting Albert Einstein's death to the class in any way compared with your father possessing his brain! "Why," she thought, "did I ever open my mouth?"

❄

Thomas Stoltz Harvey once thought about treading the same path as his father, Thomas P. Harvey, a lawyer and loquacious tobacco smoker with observations about the world ready for anyone who would listen. Harvey Sr. worked for the Travelers Insurance Company, an old American firm that had sprouted offices clear across the union. He and his wife, Frances Stoltz, were living near the branch in Louisville, Kentucky, when Tom was born on October 10, 1912. His sister, Jean, followed a year and a half later. Then the firm moved the family back to his father's hometown of Indianapolis, where the Harveys had lived and prospered for three generations. In the evenings after work, Harvey Sr. used to roll his own cigarettes and regale young Tom with family lore and stories about their proud ties to the earliest Quakers of America, exposing him to the beliefs and philosophies that would shape much of his life.

George Fox, the seventeenth-century founder of the Quaker movement, felt that the Protestant clergy, caught up in its own ecclesiastic power and the stone opulence of cathedral-building, had strayed too far from the essence of the faith of the first Christians. Fox preached that an element of God's spirit existed in every human being, so that no one need be bound by the trappings of priest and church to get closer to the Lord. The final authority for faithful souls, he claimed, was their own inner light: God's voice, which should not be confused with conscience or reason, speaking directly to them at any place, at any time. Quakers gathered in unadorned meeting halls, usually without sermon or pastor, waiting to hear the Lord. No one was expected to follow a liturgy or sing. Only if the spirit happened to move them did one of them stand to speak. Otherwise, they could pass an entire hour without a word, sometimes several Sundays in a row. Silence and patience were considered traits of a good Quaker.

Officially, worshippers called themselves the Religious Society of Friends. But because they were said to tremble in God's presence, it was their detractors who first called them Quakers. They practised no sacraments or rituals, formalized no particular doctrine or creed and promoted no specific concept of heaven, hell or any afterlife at all. This life, they stressed, was the one that counted. None of this sat well with church officials, who accused the Quakers of blasphemy, or lawmakers, who questioned their loyalty to the crown. Quakers refused to swear oaths of allegiance or fight any human—even to defend the king. They believed the lives of enemy and friend alike contained God's inner light. And if they swore an oath to the crown, that would imply that there were two standards of truth—one for officialdom and another for ordinary life. To a Quaker there could be no distinction between the sanctity of an oath and a promise to a neighbour. Their regard for the truth was so inflexible that Quaker merchants would not bargain. Yet with their reputation for honesty they built corporate successes like Barclay's Bank and the chocolate empire of Cadbury's. They lived simply, dressed in plain clothes and avoided luxuries. Worldy possessions only distracted from the inner light, eating up resources they could donate to the needy. They preached humility, tolerance and self-reliance.

After years of being arrested, whipped and jailed for their views, boatloads of Quakers crossed the Atlantic in search of religious freedom in the late 1600s. The Harveys sailed from Britain in 1706, Tom's father told him, shortly after the English Quaker leader William Penn himself. They settled first in Chester County, Pennsylvania, the state Mr. Penn founded as a refuge for Quakers and other religious minorities. From there, they farmed the Ohio River Valley before migrating to the bustle of Indianapolis. Tom's great-grandfather delivered babies there. His grandfather, Judge Lawson Harvey, sat on the Supreme Court of Indiana, and Grandmother Harvey ran a baking company that boasted the state's first "continuous motion oven" after the Civil War. Young Tom clearly had sizable shoes to fill. But if he got a good education, which his father stressed, and listened to his own inner light, as the faith prescribed, he'd discover what purpose God had in store for him.

Tom was twelve when his father decided he was old enough to attend the Quakers' Sunday Friends meetings. Travelers Insurance had transferred Harvey Sr. to Philadelphia by that time, and the family lived about twenty-five kilometres away in Swarthmore, a Quaker

town named for the English hall where George Fox used to preach.
Tom was slightly built, pale and thin lipped, with a hawkish nose and
eyes the colour of a cloudless spring sky. His temperament disposed
him perfectly to private communications with the Lord. Other
children might have grumbled or fidgeted while their parents waited
for inspirational light, but not Tom. "The silent meetings fit my
personality pretty well," he'd say. "There was no regulation on age
if a child wanted to speak, but I never did." Even when the meetings
turned to discussions of social issues or community work, as they did
once every month, Tom listened attentively and grew familiar with
the Quaker method of conducting business. He watched fellow
Quakers abstain from contradicting one another, or defending themselves or responding with anything that might be mistaken for debate.
A Quaker Friend said his or her piece without pontificating or
filibustering, and the others reflected quietly until the clerk called
on another Friend to speak.

Though women were equally welcomed, Tom's mother never
attended the Quaker meetings. She was the daughter of a Presbyterian
minister, and so, on alternate Sundays, Tom skipped the silent gath-
erings for Sunday school and church services. But if it was his father
that most influenced his religious leanings, it was from his mother
that he inherited his work ethic. Frances Stoltz had grown up the fifth
of six children in Ottumwa, Iowa, in a pillared house behind the
church where her father preached. The Reverend Stoltz died young
of a streptococcal infection, and his children had to take to the fields
to raise cows and chickens and plant corn so they could eat.

When Tom turned thirteen, his mother put him to work, plying
a patch large enough for an asparagus bed to ripen in spring, straw-
berries in June, lettuce, tomatoes and peas the rest of the summer.
He harvested a corn crop, too, and sold his produce door to door in
Swarthmore. Just what he did with his earnings, Tom Harvey could
not quite remember. Money was never terribly important to him.

In 1927, Travelers Insurance shifted Harvey Sr. back to head office
in Hartford, Connecticut, where Tom enrolled at West Hartford High
School. He studied hard, but by his own admission he never distinguished himself where academics were concerned. "I was just an average student," he would say. "Intelligent enough, but just average." Still,
he was accepted as an undergraduate at Yale University in 1930, and
performed well enough on the board exams in 1934 to gain entrance
to its prestigious medical school, which was somewhat unexpected.
He had majored in economics, intending to join Travelers and work
with his father after graduation. But in the midst of the Great Depression,
few such jobs existed. Harvey Sr.'s position was secure, but he realized
he was one of the lucky ones. "Go to medical school," he told Tom.

"It's in your blood," said his Aunt Jeanette from Indianapolis. "Your
great-grandfather was a physician."

By now Tom had grown into a fetching young man with watchful
blue eyes and a chiselled jawline. He flourished at Yale, where towering trees shaded the campus courtyards and medieval-style buildings. Tom earned a letter running the quarter mile, got elected house
manager at the Phi Gamma Delta fraternity, played tennis and waltzed
with the nurses in training. At school, he kept his marks strong in
core subjects like biology and chemistry, and found some of the less
conventional subjects fascinating as well, particularly the class Harry
Zimmerman taught on the anatomy of the human brain. Some students felt comfortable calling their professor "Zimmy," because he
was only in his thirties, but Tom addressed him formally because it
was just good manners. Zimmerman was a Yale graduate himself,
class of '27. He had specialized in neuropathology, urged on by his
academic superiors, some of whom no doubt worried that American
endeavours in the field might fall behind those of the Europeans,
notably the Germans.

In the late-nineteenth and early-twentieth centuries, German
neuroscientists ranked at the top of the heap. Having mastered the

production of sophisticated microscopes, they had then made great strides in describing the cellular anatomy of the brain, its complex networks of nerve fibres, the composition of myelin sheaths. It was a German, Wilhelm von Weldeyer, who first coined the term "neuron" in 1891, from the Greek word for sinew or cord. German psychiatrist Franz Nissl concocted the standard stain that bore his name, a violet and dark-blue dye that enabled scientists to distinguish the architecture of neurons and glial cells. In 1924, using a micro-electrode less than a tenth of the diameter of a human hair, Germany's Hans Berger was the first to record the rat-tat-tat of the brain's electrical activity, a sound that reverberated to capture the imagination of scientists around the world. North America's bright lights in the field, like Harry Zimmerman, were sent to Germany to study. On his return, Zimmerman established Yale's first neuropathology department. He often described to his students the colourful events shaping brain science abroad, which is how Tom Harvey came to hear, and never forget, the exploits of German neurologist Oskar Vogt and his French scientist wife, Cecile.

The Vogts were proponents of the idea that certain areas in the outer layer of the human brain controlled certain functions. Just as a map of Canada is divided into provinces, they believed that the cortex could be divided into geographic regions, each one in charge of producing a particular activity, like hearing, singing, adding or spelling. This was an idea popularized by the Victorian era's practice of phrenology— the first system of assigning a single cerebral address to a human behaviour. Founded by German anatomist Franz Joseph Gall, whose own brain would eventually become the object of scientific study, phrenology held that dominant traits or talents enlarged certain bits of the brain, which pushed against the cranium, creating bumps along the surface of the skull. With their bare palms and fingertips, phrenologists fondled heads, linking the elevations and indentations they felt with maps that handily laid out the location of more than twenty-seven

skills and characteristics. A protruding forehead, for example, supposedly implied benevolence. Other bumps signified secretiveness
or wit, curiosity or determination. Phrenologists traipsed the countryside reading heads in the same way fortune tellers read palms,
emptying pockets as they diagnosed people as hardworking or honest—particularly useful information for gullible customers hiring
a housekeeper or choosing a husband. But with its fickle interpretations, changing charts and charlatans, phrenology was dismissed
by the scientific establishment of the mid-nineteenth century as
ridiculous.

Efforts to "localize" abilities to distinct brain areas might have been
abandoned entirely, except that in 1861 a prominent French neurologist discovered something remarkable. Working at a hospital in
Paris, Paul Broca announced that he had pinpointed the swatch of
brain tissue that produces speech. He based his finding on a single
patient nicknamed Tan, because it was the only word he could utter.
Examining Tan's post-mortem brain, Broca found a large lesion in
his left frontal lobe, and since the patient seemed normal in every
other way Broca concluded that this particular site (later known as
Broca's area) controlled the ability to articulate language. Unlike the
phrenologists, Broca had pathological proof in the form of a pickled
brain. And so localizing—though debate about it would rage on—
was back in vogue.

By the turn of the century, scientists like the Vogts were trying
to link behaviour not to cerebral bulges, but to the microscopic
mysteries of the brain. Advances in understanding the look and
mechanics of individual neurons convinced the Vogts that the brain
was like a quilt, with different types of cells in each patch. In 1919,
Cecile Vogt, one of the first women in neuroscience, described the
cellular design of more than two hundred such cortical patches.
Meanwhile, Oskar Vogt, director of the Kaiser Wilhelm Brain
Institute in Berlin, collected the brains of people especially gifted

in a particular field and claimed he could find cellular evidence to explain their talents.

By the mid-1920s, the Vogts' work attracted world attention. In 1926, Josef Stalin hired Oskar Vogt to examine the brain of Vladimir Ilyich Lenin. Two months after the fifty-four-year-old Bolshevik revolutionary died in January 1924, the Communist Party announced that it had "decided to take all measures available in current science to preserve the body for as long as possible." Saving earthly remains seemed fashionable at the time. Tutankhamen's three-thousand-year-old mummy, dug up with its gilded sarcophagus in 1922, still lingered in the headlines. The Soviets, meanwhile, saw a chance to immortalize their ideological father and offer their religion-deprived masses a new icon, impervious to decay, like the hallowed corpses of certain saints. A mummification team toiled furiously for four months to halt the putrefaction of Lenin's corpse. They bathed it in potassium acetate and glycerine, bleached discolourations, stitched its mouth and eyelids shut. But his cranium they sawed open. Convinced the biology of Lenin's brain would bolster his godlike status, they soaked it in formaldehyde, and two years later they made the remarkable post-war move of entrusting a German to examine it. So off Vogt went to Moscow, Zimmerman told his class, to see what he could see in the pickled brain of Lenin.

Seizing a building from an American business, the Soviets established an entire institute devoted to the job. Vogt imported his fine German microscopes, dye recipes and a glinting microtome machine to slice Lenin's brain into 31,000 wafers. In 1929, he reported that he had glimpsed the physical source of political brilliance. The pyramid-shaped neurons told the story, Vogt felt. They were so big and numerous that they connected patches of the brain "otherwise widely separated," colluding to form intricate patterns. This explained "the multiplicity of his ideas, together with the wide range and the rapidity of his powers of conception, [which] produced in Lenin unusual powers of intuition," Vogt concluded. "Thus the key to a materialistic view of Lenin's genius has been found."

The pronouncements seemed outlandish to many scientists at the time. Interest in linking biology to performance waned with the rise of psychiatrists like Sigmund Freud, Ivan Pavlov and later B. F. Skinner, whose research focused attention on the power of experience and environment to shape behaviour. In the "nature versus nurture" debate, the early decades of the twentieth century belonged to the nurture crowd. Someone like Oskar Vogt, who was known to march up to conference podiums and declare that genius could be detected in a single neuron, cracked them up. Zimmerman probably chuckled at the notion himself. But he told his students that neurology was moving in from the fringes, and he predicted that the future gadgets of science would one day reveal a wondrous world between human ears.

All this intrigued Tom Harvey, but he had already made up his mind that brains were not his calling. As much as he hoped to do interesting work, a Quaker's ambitions had to be tempered with a degree of selflessness in service to others, and though he had left home for the fraternity hall he had not left the Lord: "I used to go to the college chapel on Sundays," he said. "Then I started going to the daily services. Sometimes I'd be the only one attending." Harvey planned to practise pediatrics. His professors told him he had a gentle touch. Nothing about it attracted him especially— "Sure, I liked children well enough"—but it seemed to him a noble profession. What could be more virtuous than healing babies? He planned his thesis on "The Rolling Behaviour of Infants" and thought after graduation he would join a hospital or run his own practice in Connecticut, or maybe back in the Midwest. So it was a hard lesson to learn that a man does not necessarily choose his own destiny. Worse, he never saw it coming.

In the summer of 1937, Harvey and three friends took a bus to Montreal and then hopped an ocean freighter up the St. Lawrence to Europe. After cycling across the continent, he returned at his physical peak, biking sixty-five kilometres to Hartford and back to visit his parents on weekends. He could lick a one-way trip in two hours if he felt like it. But then tuberculosis felled him just as it did ten of his fifty classmates, who withered and sputtered and dropped out of school. In 1939, the year he was to have graduated, Tom Harvey was home in bed, drinking chicken soup, disappointment crushing his chest like the fluid doctors drained from his lungs. His mother nursed him for six months before sending him to a sanatorium. He was twenty-six years old—the very age at which a fledgling physicist had made his immortal mark on the world.

Not far from the shore in New Haven County, at the end of a dirt lane twisting through brush, the eighty-hectare compound of the Gaylord Sanatorium seemed grand at first. A Tudor-style mansion housed the administration. Pines and rose arbours dotted the fields, and a farm operation sat on the perimeter, producing enough milk and meat to support the Gaylord mantra that getting fat was a sign of health. Cooks served baked potatoes for breakfast. On his arrival, Harvey realized, hacking and horrified, that most of the patients lived in tents all year round. Gaylord's other mantra was that fresh air fought the infection. Patients wrapped themselves in blankets, flimsy against the night winds, stuffed hot water jugs into their cots and wound their legs around them, waking to see orderlies balancing meal trays on their heads and goose-stepping over snow drifts. "It could be twenty below," Harvey said, "and we'd still be living in these tents." If he ever had a crisis of faith, it was there, wheezing and shivering in a city of canvas peaks, wondering whether God had deserted him. Religion made no sense to a man young and sick, scared he might have no future at all. "I was a good Protestant too until I came down with TB, and then I had all this time to think about religion," Harvey said. "I started thinking that people said God made man in His own image, but I think that man made God in his own image." It was then he decided that he was more of a Quaker, with his own personal definition of a higher, divine power, than he was a Presbyterian, with a preordained image of God.

Some patients wove baskets and played cribbage at Gaylord's recreation centre, but Harvey mostly spent his days reading. Other medical students and young doctors recovering at the sanatorium shared subscriptions to the scientific literature, and Harvey's favourite was *The New England Journal of Medicine.* He felt less sidelined knowing what questions occupied the minds of doctors, imagining the scientific puzzles he might solve, the medical contributions he might make. Gaylord staff tried to boost the residents' spirits by touting the accomplishments of former patients they had healed—among them Yale football star Albie Booth and playwright Eugene O'Neill, who won the 1936 Nobel Prize for literature.

The highlight of his stay was chatting with a fellow patient, Elouise Shawkey, a fine-boned brunette. Elouise was a nursing student he had met once at Yale, not the giggling sort who beamed at him all the time. She sported a more serious disposition, wrote short stories and poetry, and spoke her mind. Elouise was three years older than he was and she had already finished her studies to be a librarian when she took up nursing. Harvey respected the fact that she came from a West Virginia family that had graduated two doctors already, and that her father happened to be a pediatrician. Sometimes they took their prescribed walks together out past the Tudor mansion and the covered fieldstone gateway where the Gaylord bell hung. The staff rang it whenever a patient was discharged. A year and a half passed before it rang for Tom, and a short while after it would be Elouise's turn.

"Take it easy," the doctor told him when he left. "A regular medical residency program is out of the question. Try pathology, you don't have to get up nights."

Tom Harvey married Elouise Shawkey at a small chapel in the spring of 1941, the same year he finally graduated, the same year the Japanese bombed Pearl Harbor and America joined the war. Harvey had expected to regain his strength in Yale's one-year pathology program and then return to pediatrics, but Uncle Sam decided otherwise. While boys and men were being blown to bits on European shores, the military discovered that gastroenteritis and other contagious infections bred in the filth of warfare were as great a threat to manpower as the Axis enemy. The country screamed for medics, and for pathologists particularly.

Nothing did more to highlight the need to understand how the body responds to stress and disease than the war. The science of pathology predates the Middle Ages, but it was barely established as a separate field before the 1930s. General internists and surgeons practised it on the side. Only a fledgling minority promoted it as a specialty, touting the value of regular autopsies, microscopic studies and lab tests to better diagnose and treat patients. As Washington poured money into medical research, it exercised its authority to pull student pathologists into the armed forces. So it was Milton Winternitz, head of Yale's pathology department, who handed Harvey his assignment. He never did finish his residency training.

Decked in his civvies, which was just as well for a pacifist Quaker, Dr. Thomas Harvey reported to the U.S. Chemical Warfare Service at the Edgewood Arsenal in Maryland in 1943. So many of his school friends had signed up to soldier, and Harvey was glad to at least do his part for the Allied effort. He cut open rabbits, dogs, cats and rodents exposed to noxious gases and studied the pulmonary havoc that phosgene, lewisite and mustard gas can wreak. Meanwhile, scientists at another military lab in a secret, dusty patch of New Mexico were constructing the mother of all weapons. Harvey learned about it with the rest of the world after the atomic bombs blew Japan out of the war. Newspapers reported that the victory stemmed from theories that Albert Einstein had formulated thirty-five years earlier, but the physicist deeply regretted the human suffering it caused. Although he had spurred its American invention with a letter to Franklin Roosevelt in 1939, Einstein shunned any credit for production of the atom bomb and fought a long campaign against nuclear arms.

With the war over, Harvey thought about little else but getting his career back on track. Except now he had more than himself to worry about. He and Elouise had had three sons in the previous six years, Thomas in 1942, Arthur in 1944, and Robert in 1946. In 1948, the family packed up the small military house at Edgewood and left for Pennsylvania, where Harvey's ancestors had first settled and where he hoped to finally become an intern at the Philadelphia General Hospital. But when they arrived, he discovered that the hospital paid medical residents only $50 a year, hardly enough to feed and house Elouise and the boys. "I figured I'd better go back into pathology, because I could earn a living all right in pathology," Harvey said. "You could make $4,500 a year at least, and that was pretty good. So that was what I did, I went back."

The University of Pennsylvania hired Harvey as an assistant to neuropathologist Frederick Lewy in 1948. Seven years would pass before Harvey considered it an auspicious appointment. Lewy, a small Jewish man in his sixties, had a friendly nature and a thick German accent. He had headed the Neurological Institute in Berlin before the Nazis turned him into a refugee. The German connection reminded Harvey of Harry Zimmerman's lectures and he wondered if Lewy had ever met Oskar Vogt, but he never did ask. After all, Lewy was famous in his own right. In 1914, he had discovered that patients who suffered neurodegeneration, particularly Parkinson's disease, appeared to have unique cellular structures in their post-mortem brains. Elongated blobs of plasma hid in the cavities of their neurons, in the hypothalamus and substantia nigra, the nub of tissue deep in the brain that produces dopamine and controls voluntary movement. He described these unusual masses in the medical literature in 1920 and they became known as Lewy Bodies, a sign of neural deterioration. It impressed Harvey to work lab coat to lab coat with Dr. Lewy of Lewy Body fame.

Lewy believed in providing his assistants with all sorts of opportunities; he had even brought some of his Berlin staff with him to Philadelphia. At his urging, Harvey took a few courses in which he learned how to prepare brain tissue for microscopic slides, and he taught Lewy's neuroanatomy classes for two years. But when Lewy died in 1950, Harvey left brain studies for a job in clinical pathology at Philadelphia General Hospital. He had not given up his hope of working with patients, even if they were dead ones. Recent breakthroughs like the vaccines for diphtheria and smallpox, had elevated the status of pathologists. And the truth was that pathology fit Harvey like a pair of latex gloves. He could work alone, in utter silence if he wished, in a profession that he was coming to believe was as noble as any other. Doctors relied on his lab work for their diagnoses. "I especially enjoyed doing autopsies; you can learn a lot and you're in a teaching position then at the hospital because you can explain all these things to the staff," he said. "I saw people die and the doctors didn't know why and I would find out why." His enthusiasm did not go unnoticed. When a fellow named Jack Kauffman telephoned Philadelphia General in 1951 looking for someone to head the new pathology department at Princeton Hospital, Harvey's boss did not hesitate to recommend him. Harvey rushed home to tell Elouise that they would be moving, and she said she was anxious to go. The pace and noise of Philadelphia frayed her nerves. Princeton was a quiet little college town, Harvey told her, full of scholars and important people.

✦

To the children of Princeton, Albert Einstein was the gentle genius who lounged on his front porch in gigantic fuzzy slippers smoking a pipe. He helped them with their homework and serenaded them on his violin when they trick-or-treated at Hallowe'en. They never gawked at him if they saw him strolling down Nassau Street in his tennis shoes, slurping an ice cream cone dipped in chocolate and candy sprinkles. Embarrassed adults would have glared at them disapprovingly if they had dared. Princeton teemed with famous people, and it would not do to swoon like schoolgirls at a Frank Sinatra film every time you went to town. Jimmy Stewart and F. Scott Fitzgerald were Princeton alumni; Paul Robeson of "Ol' Man River" fame grew up on Witherspoon Street; Upton Sinclair, one of Einstein's favorite authors, once wrote nearby; Dr. George Gallup, of the world-renowned Gallup polls, called the town home, and so did many other big names besides. It was nothing for a child to hear a friend couldn't come out to play because novelist Pearl Buck was coming for lunch. The children knew that they lived in a town where things happened, and had happened for three hundred long years.

Quaker farmers first settled the region's dense woodlands in 1696, and the stagecoach brought business and people. In 1777, during the War of Independence, George Washington's army waged a bloody skirmish on a Princeton farm. At the time, Princeton University was a single stone building with a soaring steeple named Nassau Hall, in honour of England's Prince of Orange-Nassau. That the British tried and failed to burn it down only launched the school's enduring reputation for producing champions. Princeton attracted students from as far away as the West Indies and graduated two presidents. Around it the town grew. Princeton's forefathers imitated the architectural styles of other cities and lands so that parts of it resembled a New England village, or a British high street, replete with towers and gables and stained glass, Gothic, Dutch and Tudor Revival. Foreigners admired its genteel charm and European ambiance. When Einstein arrived in 1933, he wrote to his good friend the Queen of Belgium that "Princeton is a wonderful little spot, a quaint and ceremonious village."

Princetonians worked hard to preserve the past. They banned the blight of neon signs and anything resembling a skyscraper. Saving hallowed landmarks from natural decay was a local preoccupation. They shuffled old buildings off their foundations to more convenient locations rather than levelling them. Engineers constructed an actual field of railway tracks on which to mount the original Woodrow Wilson School of International Studies and push it back a mere 75 yards to make room for an expansion. Naturally, they fretted over the post-war technology boom. On nearby Route One, the Radio Corporation of America hired new workers to tinker with the first colour televisions. South of there, a factory had started mass producing the miracle drug penicillin. At the Princeton Plasma Physics Laboratory, employees experimented with energy alternatives to radioactive nuclear fission. Scores of engineers and scientists arrived with their families while prefab houses sprang up where corn husks used to stand. Surrounding townships swelled like doughnut batter in a deep fryer. Editorial writers at the *Princeton Packet* newspaper declared their sympathy with "those who sincerely hate to see the town get any bigger." But Jack Kauffman was delighted. Princeton Hospital's visionary administrator figured that if the town grew, the hospital could shake off its ragtag beginnings and grow right along with it.

Until the 1918 worldwide flu epidemic, most residents saw no need for a town hospital. When an earlier attempt had been made to build one at the corner of Nassau and Princeton Avenue, violent protests had erupted in the neighbourhood. Princetonians preferred to seek their doctoring in New York or Philadelphia, and nearby farmers were reluctant to hobble in to the snooty town with their broken bones and muddy shoes. So the hospital, which eventually sprang up on two donated hectares where dairy cows once roamed, was not even a poor cousin to the university up the road. It took a clatch of wealthy and powerful trustees who banded together during the 1930s and '40s to turn the three-storey brick building into a proper medical institution with well-trained doctors, beds for emergencies and a nurses' residence. People like George B. Lambert, who marketed Listerine and blades for the Gillette Razor Company, and Hack McGraw, of the publishing empire, raised money, encouraged charity drives and women's auxiliaries and hired the bulldog Jack Kauffman as their agent of change.

Kauffman had the scrappy determination that a man who never went to medical school can cultivate when he finds himself in charge of a bunch of doctors. He smoked two packs of cigarettes a day, ordered the daily scrubbing of hospital corridors and reviewed the protocol of every department. He convinced the trustees to hire a consultant from the American Hospital Association to study the organization of medical staff and their relationship to the community. When the consultant's report landed in 1950, the recommendations were clear: hire more specialists; do not allow doctors trained in one field to practise in another; create departments for every specialty; and, specifically, recruit a full-time pathologist to examine surgically removed tissues and determine causes of death by performing more autopsies. Up to that point, the hospital's chief surgeon had done the pathology work, reviewing tissues removed in his own surgical procedures.

Kauffman also crafted a clever plan to lure patients from every stratum of the community. He knocked down walls to add an outpatient clinic, where he hoped everyone, from high society on down, would visit their doctors, dentists and obstetricians. That way, patients who came for a check-up might be enticed to take advantage of the hospital's other services should they need an X-ray or an operation. Kauffman renovated with his characteristic tenacity, roping in New York DJs for fundraising parties and raffling off Fords and Thunderbirds to finance construction of two adjoining office towers.

On one of the upper floors, Henry Abrams examined eyes. Abrams was just the sort Kauffman was after for the new digs. The young ophthalmologist held free eye clinics for the needy at the Lions Club, and yet he had also collected a patient list that included Albert Einstein, Princeton's most famous resident. So it was a coup that Kauffman had convinced Abrams to relocate his Nassau Street prac-tice to the hospital on Witherspoon—and, by all means, to bring his patients with him.

Fresh out of medical school at age twenty-six, Abrams had opened his general practice in downtown Princeton in 1939. For two years, he drummed his desk waiting for patients, and then one day, just like that, the family doctor who worked next door asked him a favour. Conway Hiden told him that he'd decided to concentrate on a sur-gical career, and he wanted to know if young Abrams would take over his practice. "When I heard Albert Einstein would be among my new patients, I nearly fainted," Abrams said. The rapture never left him. No matter what restrictions adults imposed on the reactions of chil-dren, even celebrities fell dumbstruck near Einstein. After visiting the physicist at his Princeton home, novelist William Faulkner described Einstein as exuding a mystical quality that left him speech-less. Artist Winifred Rieber, who painted Einstein's portrait, said, "Everything about him is electric. Even his silences are charged." Abrams described being near Einstein as nothing less than a religious experience: "I felt like I was flying in heaven."

Helen Dukas, who had been Einstein's devoted housekeeper and assistant from the time she was hired in Berlin in 1928, sometimes telephoned Abrams and invited him to visit. "Henerry," she'd say in her German accent, "ze professor vould like you to come over and talk." Abrams always went, of course, and sat with Einstein in his small front room, talking about nothing in particular—not physics or "any deep matter." Sometimes they discussed politics or Abrams's plans for the future, and the young doctor concluded that it must do Einstein good to play father figure to him. He had been reading biographies about his famous new friend and learned that Einstein's younger son, Eduard, had been diagnosed with schizophrenia and committed to a psychiatric hospital in Switzerland. And though the physicist's elder son, Hans Albert, had arrived with his family in America in 1938, father and son rarely saw each other. "The professor carried guilt, I think, that he had not been able to spend time with his own sons, and perhaps he regarded me like a son," Abrams later said. They wrote to each other while Abrams served in Greenland during the war, and when he returned their friendship deepened. Einstein became godfather to Abrams's son Mark, and it was Einstein who encouraged Abrams to specialize in ophthalmology. But the physicist forbade Abrams to schedule private eye appointments for him when he opened his new practice at Princeton Hospital. Einstein preferred to sit in the waiting room until his name was called, just like everybody else.

Every year Abrams peered into Einstein's coffee-brown pupils and reached the same conclusion: "Your vision could be improved with a pair of prescription reading glasses," he would tell him. But every year Einstein pulled out the five-and-dime magnifying eyeglasses a friend had sent from New York.

"These work well enough for my needs," he would say.

Guy Dean, a youngish doctor who slicked back his wavy hair, became Einstein's family doctor at Princeton Hospital. He had tended the physicist's younger sister, Maja, before she died of pneumonia in Einstein's Mercer Street house in 1951. The professor still retained specialists in New York, but Dean checked Einstein out regularly, told him to quit smoking his pipe, as other doctors had, and ordered tests for him at the new pathology lab.

A full-time pathologist had at last been hired from Philadelphia in 1952, and the staff was sincerely glad to welcome Tom Harvey on board. Sure, some found Harvey as bland as boiled milk, because he didn't strut or swear or tell off-colour jokes. He never even raised his voice. Mrs. Godfrey, the radiologist's wife, remembered him as the quiet one at staff parties in the hospital cafeteria. Jack Kauffman respected his qualifications but probably found him a little peculiar too. Kauffman stopped by the lab one December day and discovered the pathologist cooking his family's Christmas turkey in the autoclave, the high-tech pressure chamber normally used to sterilize the morgue instruments.

People might have chalked up any strangeness to the nature of Harvey's job, cooped up all day in the laboratory with fresh-cut tissue and the refrigerated dead. But everyone agreed that the new pathologist was exceedingly polite and had a charming smile. He impressed them as a good listener, too, so that when obstetrician Raymond Stone mentioned he had served with a neuropathologist at an army base in Puerto Rico during the war, Harvey never forgot it. When he did speak, he chose his words gingerly, and a nervous laugh punctuated his sentences like too many commas. It wasn't the sort of chuckle that would invite people to laugh with him, but it gave the impression that he was never aggressive with a viewpoint. His reticent attitude reflected his Quaker roots, but beyond that Harvey realized that showing even a hint of arrogance in his position would be politically ill advised. He understood the natural tension between pathologists and doctors. Pathologists can see what a doctor has missed: a fatty liver misdiagnosed as a bad heart; a brain tumour mistaken for a stroke. Louis Fishman, the hospital's medical internist, had great admiration for Harvey and felt he had a good deal in common with a doctor who navigated human physiology with his hands in those decades before they had fancy scanning machines. Fishman estimated that as many as 40 percent of clinical diagnoses doctors made were unsupported by the autopsy results. In those days, a good hospital cut open more than half of its cadavers, he said, the general sense being that pathology had rescued medicine from the Dark Ages. As for Harvey, he "felt lucky to find that situation at Princeton, a small ninety-bed hospital that had had no pathologist. It was fun."

Harvey and his wife bought a house a few blocks from the hospital at 245 Jefferson Road. It had belonged to the French chef who had cooked for university president Woodrow Wilson before he became president of the United States. Other doctors lived nearby, as well as a banker, a university Latin professor and a woman widowed by the war. The leafy neighbourhood might have been considered the posh side of town in any other place, but not in Princeton, where the heirs to the Johnson & Johnson fortune owned enough land for a golf course and New York tycoons donned their slippers in eighteenth-century mansions. Yet by Harvey's standards it was grand, a two-storey Edwardian, covered in a mustard-coloured stucco with shutters and dormer peaks and enough green space surrounding it for the boys to hurl a football without breaking a window. Robert, his youngest, was seven when they moved in, and the boy felt a pang of privilege seeing the house from the outside for the first time. "But I don't ever remember living extravagantly, and I never had the feeling my family was affluent," he said. "I wore my older brothers' clothes." He swore they must have been the last family in the neighbourhood to buy a television, the last kids on the street to tune in to *The Mickey Mouse Club* and *Davey Crockett*. With all its non-stop talking, his father thought television "was rubbish."

The house boasted a formal dining room but the Harveys rarely ate there. They sat in the kitchen, at a wooden picnic table. One of the benches sometimes collapsed during meals and spilled the boys on to the kitchen floor. Harvey would then haul his toolbox up from the basement and nail it back together. On the porch, Harvey still kept the three-speed Raleigh he rode through Europe. In the driveway, they had a 1937 DeSoto, which they drove until it died, and a 1952 Ford. Every summer, when Princeton faculty and students deserted the town, Harvey would load Elouise and the boys into the Ford and drive across the country. He would take a full month off, cram the car with tents from L. L. Bean and a cooler full of wieners and clock eight hundred kilometres in a day. Harvey mapped out the trips using his collection of travel guides, camping some nights only to pack up at first light and hammer the road again. They toured the Great Smokies, the Grand Canyon, Shenandoah, Glacier and Yellowstone parks and the rocky shoreline of the Canadian Maritimes. His sons relished their summer holidays. The rest of the year their father was a very busy man.

Harvey never mingled on the cocktail circuit; civic obligations filled too many of his evenings. Social action to improve the lives of others is a cornerstone of Quakerism, and he considered it his responsibility to contribute to charitable and progressive causes. He joined the Individual Liberties Association, sponsored by the League of Women Voters, which was mulling over the Bill of Rights and how it applied it to local issues—like the new town swimming pool and the reluctance of some to allow blacks and whites to swim together. The Rotary Club elected Harvey president of the Princeton chapter to spearhead its humanitarian and community projects. He even sat on the board of the local Planned Parenthood office, a bold venture in 1950s America. But nothing about testing Pap smears or advocating birth control clashed with his faith. The Society of Friends treats issues surrounding sexuality as an individual choice, guided by conscience. To Harvey, birth control was preferable to bringing unwanted children into the world.

At the hospital, Harvey established a Tissue Committee and took responsibility for reviewing every gallstone, appendix, cyst and any other bit of flesh removed in surgery. He sliced and studied the specimens under a microscope in his three-room laboratory and determined whether the operations had been necessary. He also presided over the hospital's monthly Mortality and Morbidity conference, at which staff discussed interesting cases and how doctors might have handled patients differently given the full knowledge of the autopsies.

Before long, Harvey found himself conducting dozens of laboratory tests every week and performing nearly one hundred autopsies a year. With a nod from a happy Kauffman, he decided the workload demanded extra staff. The Council on Education of the American Medical Association approved Harvey's request to start his own training school for lab technicians at Princeton Hospital. Young women, not necessarily inclined toward nursing, ended up as students in Harvey's lab. Elouise Harvey kept clippings of early enrollees when the local paper ran their photos: her husband, wearing a bow tie, hovering behind a Miss Harriet Honda as she looks down at the freezing microtome capable of slicing deli-thin sections of tissue; a Miss Dina Baldasaria sorting vials in the chemistry lab. The young techs also ran errands for Harvey and made house calls to collect blood or other bodily fluids from Princeton patients. When Einstein's doctor, Guy Dean, passed along his requests, it was the young women Harvey usually sent to the physicist's house on Mercer Street. They always returned to the lab in good spirits, completely charmed by Einstein's wit. "The professor is a wonderful man," they used to tell him. So one day when Dean asked, Harvey decided to go himself.

He nosed out on to Witherspoon Street, turned left on to Nassau, past the university, hung a right on Mercer and pulled up in front of number 112. It was nothing fancy. A narrow one-and-a-half storey, painted yellow except for the hunter-green shutters that flanked the windows. It had actually been moved to Mercer from a neighbouring street in 1876. Harvey had of course driven past it several times since moving to Princeton, but now here he was, preparing to go inside. He climbed the porch steps where Einstein liked to sit, carrying his black leather medical bag. Helen Dukas, a pint-sized woman of boundless efficiency, answered the door as she usually did, ever ready to protect the professor from an overzealous press or adoring strangers. Einstein affectionately called her his Cerberus, the three-headed dog of Greek mythology that guards the entrance to the Underworld. Harvey smiled, lifted his medical bag and told her Dr. Dean had sent him.

Dukas showed him upstairs past the front room sparsely furnished with heavy German dark wood. A violin case leaned in the corner. Einstein was clearly a modest-living man, Harvey thought. Dukas opened the door to the professor's bedroom and ushered Harvey in. "It wasn't a very big room," he said. "The bed took up half of it." And there he lay, under an eiderdown, his fabled mane flattened against a pillow.

"I see you've changed your sex," teased Einstein, who no doubt preferred the company of the youthful female lab techs. Harvey laughed sincerely and introduced himself as Princeton's chief pathologist. Having the opportunity, another man might have tried to strike up a conversation with Albert Einstein, told him about cycling through Germany, past concentration camps and Panzer tanks amassing. But Harvey said nothing. "I regarded him like everyone did," Harvey said, "as the genius of the age."

Einstein was frequently plagued by congestion and Harvey thought he was likely suffering from a garden-variety respiratory infection. He rolled up the sleeve of Einstein's bathrobe and stood over the professor in silence as he jabbed the syringe into his arm. He swabbed his pricked flesh with a cotton ball and taped a bandage across it. Then he gave the genius a jar and the professor disappeared.

Harvey thanked him when he returned a few moments later and said Dr. Dean would be in touch with the results. "He was very friendly, very easy to talk to for someone who had such a big presence in the town," Harvey remembered. He left that day with Einstein's blood and urine samples. It would not be the last time they met, but it was the only occasion on which words passed between them.

THREE

'55-33

✺

EINSTEIN REFUSED TO come out on what would be his final birthday. Dozens of camera and newsreel men waited on Mercer Street, as they did every March 14, passing the time by snapping photos of each other in front of his house. Helen Dukas stepped onto the porch at 9:00 a.m. "No pictures today," she told them. "Dr. Einstein says use last year's picture. He hasn't changed since then!"

Some time later, a car pulled up while the newsmen were still gathered, hoping Einstein would change his mind. Two men got out, Benjamin Cortizano, an artist from New York City, and his friend. They carried a large package to the door, and Dukas ushered them inside. When they emerged a while later their car wouldn't start. Cortizano asked a Hearst Metrotone News photographer named Nicholas Archer for a ride uptown. When Archer dropped him off, Cortizano repaid the favour with a roll of film. He said it contained pictures of Einstein taken that very day with the professor's own camera. Archer rushed the film to the newspaper office of the *Princeton Packet* and asked if someone could develop it. Then he happily phoned his editors in New York to report his coup. He had them, he said, the only pictures of Einstein at seventy-six.

His editors, Archer said, "turned a flip!"

In one of the photographs, Einstein is leaning against the railing of his back porch, bathed in a sharp wedge of sunlight. Cortizano stands beside him presenting a completed portrait, but the gradation of shadow obscures the artist so that you see only Einstein, gripping the frame with his left hand and gazing downward at his painted self, as if to approve the benign image that would accompany his own obituary a month later.

Einstein knew he was running out of time. Not because of his bouts of respiratory troubles or the anemia sucking oxygen from his blood, but because of the monster growing in his belly. It had been mid December 1948 when he had first arrived at the Brooklyn Jewish Hospital clutching his middle. Doctors suspected gallbladder trouble. But surgeon Rudolph Nissen discovered something else when he cut him open: an ominous, grapefruit-sized bulge in the abdominal region of his aorta, the main artery that arches off the heart, driving blood down along the left side of the spine and forking into two descending channels that quench the lower body. Three-quarters of aortic aneurysms appear near the abdomen. Weak tissue balloons with rushing blood, threatening to explode. Nissen, who pioneered the procedure, wrapped two-thirds of the aneurysm in yellow Cellophane, hoping to keep the engorged artery intact. The doctors told Einstein afterward that it might yet burst.

"Let it burst," Einstein replied.

He left the hospital by the back entrance, poking out his tongue at the gauntlet of savvy pressmen who waited there. When the famous cheeky pose appeared in the newspapers, Einstein clipped it, penned a farewell caption and sent it to the surgeon: "To Nissen my tummy / The world my tongue."

So it was that on April 12, 1955, Einstein grimaced with a pain he had been expecting for seven years. Two days later, he collapsed on the bathroom floor. Guy Dean told Dukas to bring him straight to Princeton Hospital when she called, but Einstein refused to budge from his own bed. Dean had to settle for a house call. He wrote in his notes that day that Einstein looked pale, practically blue, and was suffering from nausea and pain in his lower abdomen and back. Dean brought an electrocardiograph and evaluated the professor's heartbeat. It was too slow. Then he pressed his hands against the tender flesh of Einstein's belly. The aneurysm had grown; it was throbbing now and leaking blood. Dean called in Frank Glenn, New York Hospital's chief surgeon, to consult. Glenn concluded that Einstein required urgent surgery. But the professor resisted. "I've lived a long time, [have] always been busy and enjoyed life, and why go to all the trouble of an operation," he said. "I want to go when I want . . . it is tasteless to prolong life artificially. I have done my share, it is time to go. I will do it elegantly."

Dean left morphine for Helen Dukas to inject. She stayed up most of the night wetting Einstein's dry mouth with mineral water and ice cubes. The next day, Friday, April 15, swayed at last by the toll that round-the-clock nursing had taken on his housekeeper, Einstein agreed to go to Princeton Hospital. According to biographer Denis Brian, he telephoned Dukas soon after he arrived there and asked her to fetch his spectacles, pens and papers. Hooked to an intravenous glucose drip and high on morphine, Einstein still struggled to find a unified field theory. Princeton doctors meanwhile conferred with Einstein's New York physician and they all agreed that the aneurysm had ruptured. In his notes, Dean wrote that the bulge had swelled "from just to the right of the midline to well out in the flank . . . the pulsations of the mass are frightening in their magnitude." Surgery was the only hope, a bold undertaking to sever the bulge and bridge the arterial gap with a plastic insert or cadaver tissue or a spare piece from another of his vascular passages. Chances of survival were as low as 50 percent, but an untreated aortic rupture would surely kill him. A parade of physicians, family and friends pleaded with Einstein to undergo the operation. Thomas Bucky, who had introduced Harry Zimmerman two years earlier, was among them. Bucky, a doctor himself, arrived from New York and urged Einstein to travel back with him to the city, to the operating theatre of a leading vascular surgeon. "We all tried, but it made no difference," Bucky said. "We couldn't change his mind."

Einstein had come to distrust medical wizardry. He had little faith that innovation had made it possible to operate on a man in his condition. He was by turns cynical and bemused with all the physicians who treated him in later life, once telling his friend Henry Abrams that it was just as well doctors failed at their tasks as often as they did. The planet, he said, would otherwise be overrun by a population explosion. Instead of contemplating how his life might be saved that April weekend, Einstein issued instructions to be followed after his death.

The fate of his remains always concerned Einstein, since people had idolized him like a living monument. Mobbed during a trip to Geneva, a crazed young girl had tried to snip off a lock of his hair. What might they do to his body when he was dead? "I want to be cremated so people don't come to worship at my bones," he had once told his friend and biographer Abraham Pais. Judaism frowns on the practice, but like his parents, Einstein held unorthodox religious views. Cremation seemed a prudent, practical end for his earthly self. What's more, Einstein told his hospital visitors, "Don't let the house become a museum." He wanted no pilgrims on Mercer Street.

Of all those who sat at his bedside, it was likely his Cerberus, Helen Dukas, and Otto Nathan, his loyal friend and adviser, that Einstein presumed would carry out his wishes. Nathan, a former financial counsellor to Germany's ill-fated Weimar Republic, had taught economics at Princeton after fleeing the Nazis and was by this time an economics professor at New York University. He had arrived in America shortly before Einstein and helped the physicist adjust to his new life in the United States. Einstein remained fiercely grateful. The vaunted professor had been homesick for the like-minded scientists and social democrats he'd left behind, and Nathan helped fill the void.

A foot soldier in the First World War, Otto Nathan was a devout pacifist and deeply passionate socialist—dangerous ideals in America's McCarthy era. But while the two men shared common political views, their personalities differed dramatically. Nathan could be prickly, even sullen with strangers. Thomas Bucky, who often discovered Nathan at Einstein's house when he went to visit, described him as a contrary, tactless fellow who went off like a shotgun at the slightest thing. "He was unflinchingly opinionated and could be impossible to deal with," Bucky said. "But he was quite influential with Einstein, even on personal matters. . . . We all questioned, 'Why the hell him?' My father [Gustav Bucky] took Einstein aside and talked to him about it, but it made no difference." Nathan's appearance seemed to speak volumes about his disposition. He viewed the world through owlish spectacles and, at five feet two inches, from the shoulder height of most other men, the shaggy hedges of his eyebrows often framing a leery glare. Yet his manner suited the mission to which he had devoted his life. Like Dukas, Nathan never married and instead committed himself to handling Einstein's daily deluge of mail, thwarting the press and shielding the professor from his enraptured public. Jamie Sayen, whose family lived next door to the Einstein house, knew that the economist came off as irascible but understood completely what Einstein must have seen in him. "Nathan was a man of honour," Sayen said. His unflagging sense of justice and decency, Sayen felt, earned Einstein's respect and trust implicitly. Nathan managed the physicist's finances (who was known to use cheques as bookmarks), supervised his personal affairs and considered himself a valued member of Einstein's unconventional family of guards and gatekeepers.

More than two days passed before anyone contacted Hans Albert in California with the news that his father had collapsed. Einstein's elder son, who taught hydraulic engineering at Berkeley, caught the first flight out. He and his father were seldom seen together, and many people were unaware that Einstein even had a son. But no one ever let Hans Albert forget that he had a father. People asked so often if he was related to the genius Einstein that Hans Albert once compared the question to the steady drip of Chinese water torture. As biographers Roger Highfield and Paul Carter put it in *The Private Lives of Albert Einstein*: "Hans Albert bore the emotional legacy of a man whom by turns he had worshipped and despised."

The boy had been only ten years old when his parents separated, and Hans Albert had simply assumed that his father, then five years away from fame, did not want the family to distract him from his work. He and his brother Eduard returned to Switzerland with their mother, Mileva Maric, a Serb and a bright physicist in her own right, leaving Einstein in Berlin. Hans Albert became the little man of the house. Ties to his father strained and snapped at various points through the years: when Hans Albert chose the science of rivers over the science of the universe; when he chose a wife his father disliked; and in 1919, when Einstein married his first cousin Elsa just four months after divorcing the boys' mother. It seemed particularly sad to Hans Albert that his father had so little contact with his institutionalized brother Eduard. After moving to America with Elsa, her daughter Margot and Helen Dukas, Einstein never again laid eyes on his younger son. Yet Hans Albert, who shared his father's love of academics, music and sailing, also inherited his distaste for personal sentiment. His classmates once called him "Little Stone," and his own daughter would remember him as a cold and distant father.

There were no reports of an emotional outpouring or tears when Hans Albert arrived in his father's hospital room, but Einstein was pleased to see him. The two men discussed science until Otto Nathan arrived, and the conversation turned to the question of whether Germany would try to rearm. Hans Albert tried, like the others, to convince his father to undergo surgery, and when he left that Sunday, April 17, he thought that with a little more time he might sway him.

Word of Einstein's poor condition spread quickly through the small hospital. Nurses shared their observations and surgeons discussed the odds of saving him, the way surgeons do when a delicate procedure demands flawless precision from their knives. D. Barton Stevens, one of the Princeton surgeons, did not blame Einstein for refusing the operation. Vascular surgery, he said, was then in its infancy: "You would not want to undergo an operation like that in those days. If you can imagine, it would be like stitching wet tissue to cottage cheese."

The futility of the discussion greeted Stevens in the hospital lobby on Monday morning, April 18. Reporters and photographers milled about as he came in to make his rounds. They stopped nurses, orderlies and doctors to quiz them for details about the death of Albert Einstein. The genius's last thoughts had been muttered in German to a nurse who did not speak the language. Einstein had then stolen two final breaths and died at 1:15 a.m. The hospital had waited seven hours to contact the media. A spokesman at the time said it was not hospital policy to take the initiative in announcing deaths, but if anyone had called to inquire about Einstein's condition, the news would have been freely released.

Dr. Guy Dean had checked on the sleeping professor at eleven o'clock Sunday night. He'd left the hospital with the vain hope that Einstein's rupture might somehow repair itself. A few hours afterwards, a nurse had telephoned him at home, then notified Jack Kauffman. Dean had then called Hans Albert, who'd sat up comforting Helen Dukas through the night. Finally, sometime around dawn, Dean had dialled the Harvey home on Jefferson Road.

✳

At school, Tom Harvey used to read the kinds of books with plots that thrust ordinary boys into extraordinary adventures. He devoured

Tom Sawyer and *Huckleberry Finn* and *Treasure Island*. Dickens was his favourite author. *A Tale of Two Cities* captured his imagination with its sweeping drama and cast: Dr. Manette, wrongly imprisoned for eighteen years; Sydney Carton, the invisible jackal; and Jerry Cruncher, digging up fresh graves to sell corpses to science. In it he encountered words of resurrection, sacrifice and emotion spat raw from the characters' mouths: "If you could say, with truth, to your own solitary heart, to-night, 'I have secured to myself the love and attachment, the gratitude or respect, of no human creature; I have won myself a tender place in no regard; I have done nothing good or serviceable to be remembered by!' your seventy-eight years would be seventy-eight heavy curses; would they not?"

They would indeed.

On the morning of April 18, Harvey buttoned up a white cotton shirt and looped a thin dark tie around his neck. At forty-two, his firm jaw line had yet to melt into the soft flesh bestowed by a good wage, and that spring morning he was an altogether happy man. He strolled up Jefferson to the hospital. Leaves budded on the trees and the grass was greening. It was a fine day. And Hans Albert had given Guy Dean permission for the hospital—for him—to perform an autopsy on his father.

He arrived early, before the media had assembled, though staff members had already parked their cars on the asphalt out front. The day was poised to begin as usual. Cooks prepared hot meals in the cafeteria. Doctors made rounds. Upstairs, smoky plumes billowed through the surgical department, where nurses "boiled up the sharps" for the day, sterilizing all flesh-handling instruments in a vat of bubbling mineral oil. Harvey's domain sat on the first floor. Morgues at ground level eased the chore of undertakers who came to wheel the bodies away. To get there from the front doors, you had to follow a corridor on the right, down a concrete hall that prized expedience over aesthetics. There was no window on the wooden door of the old pathology department, no glass to peek through. Usually, no one cared to. Off to the back inside were two rooms that served as the tissue and chemistry lab, but these went unnoticed when the door opened because something else attracted the eye: a stainless-steel table, mounted on a metal stump in the middle of the floor, glinting under stark fluorescent light.

Careful planning goes into designing an autopsy suite. At Princeton Hospital, smooth linoleum floors, simple to clean, sloped toward a central drain to catch the runoff of body fluids. Vents installed in the ceiling blew any spray toward the floor. The cool shade of mint-green tile chosen to cover the walls contrasted with the natural pigments of a fresh cadaver, making it easier to spot tissue discolourations. To the left of the doorway, on the other side of the medicine cabinet where Harvey kept supplies and his black medical bag, two sinks protruded from the wall: the shallow one stood at waist height; the second, deep enough to contain the splatter of a small intestine, reached to mid-thigh. A hose was attached to the long-necked faucet over the second sink. It ran up the side of a utility shelf through a metal casing to which another hose could be fastened and water directed where needed. Other tubes connected to barrels of embalming fluids that sat high on the shelf, relying on gravity to release their potions. All of it stood, hung and dangled in close proximity to that cold slab of a table, riddled with holes like a colander for life's liquids to strain through.

On the far wall opposite the entrance, behind one of the square steel doors, Einstein's body had chilled. But someone, at some point that morning, earlier even than Harvey had arrived, must have retrieved it from the refrigerator. Someone turned on the faucet and, using the hose, sprayed water across the table, making it slippery enough to drag Einstein's supine corpse by the left arm and leg onto it from the gurney. It lay waiting when Harvey arrived. He was not surprised to see it there, since Guy Dean had telephoned. But standing in front of the naked shell of the century's intellectual hero, a frisson of awe skittered through his veins. "I felt humbled," he later said.

According to Harvey he was not alone in the autopsy suite that morning. The formidable Otto Nathan watched. Einstein had appointed the sixty-two-year-old to be executor of his estate, and Nathan perhaps felt it was his duty to witness the most complete and final physical on his exalted friend. He never discussed what he saw with anyone, or left any written record of his having been there, but Harvey would describe him hovering in the morgue, near Einstein's famed feet, speechless. The dead professor's head hung over the table's edge, at the end closest to the door, so that the neck arched and extended, raising the chest to facilitate the work. In the hours since his death all the blood had sunk and settled to the under-side of his body, forming a perfect horizontal line of crimson along the corpse.

Harvey was all pathologist then—not father, son, husband or civic activist. He pulled his plastic apron over his head and tied a bow behind his back. He stretched his fingers into the sockets of latex gloves and snapped them over his knuckles and wrists. Then a familiar silence draped the room, hung heavy like a curtain deadening the racket in his own head as he summoned concentration. To think his life had somehow led him here . . . to think of all the false starts and stolen dreams—the tuberculosis, the war and the things you sacrifice to make a living—and there he was, an ordinary man, charged at this moment with the task of dissecting perhaps the greatest scientist who ever lived. He could hardly believe it himself. How would he describe the power of this moment in the years to come?

"I felt lucky," he would say. "I had the great fortune of being the one, at the right place at the right time. It was the biggest moment in my life."

He never expressed emotion easily, excitement least of all. Yet so many would come, strangers and Einstein fanatics, to understand a glimmer of the octane that surely fired him that morning, and he would struggle to offer a detached, professional assessment: "I realized that this was the body of a great man, but I didn't want to do anything differently. I didn't do anything differently."

From a ceiling hook above the torso Harvey suspended a grocer's scale and positioned over Einstein's stiffened knees a foot-high wooden table where he could lay the organs out. From the medicine cabinet he retrieved paper and pen, and he filled in the spaces on Princeton Hospital's standard autopsy report. Name, sex and age of deceased: Albert Einstein, male, 76; 55 for the year; 33 for its rank among the number of autopsies performed so far that year.

☼

Autopsy, from the Greek *autoptes*—it means to see with one's own eyes what history has imprinted on the body, what marks life has left. Every nick and callus, scar and bruise encrypts its own story on the shell that housed the living and all its passions, dreams and compulsions. A darkened notch on the third finger of a right hand, indented by the obsessive pressing of a pen. Arms still taut after summers of rowing small boats and drawing Mozart sonatas across the strings of a violin. Soles hardened with years of barefoot liberty, even when meeting a queen. It betrays us, too, post-mortem flesh, an archive of weakness. Lungs speckled black from the pipe he could never entirely give up. Arteries streaked yellow from fat drippings sopped up with hunks of bread in his youth and chocolate-covered ice cream cones slurped in defiance to the end. And there, a worm-like scar embossed below the diaphragm, a souvenir of the Nissen surgery that had first revealed death's steady march.

For two and a half hours with his faint Midwestern accent, Harvey recorded with cool, scientific detachment what he saw. Words like *lesion*, *contusion* and *lateral hypostasis* blunted any sentimentality for the brilliance lost to the world. It began with measurements. Height: five feet seven and a half inches (176 centimetres); Weight: 180 pounds (81.6 kilograms); Chest: thirty-four and a half inches (87.6 centimetres) across. Then Harvey turned to his trolley of tools, to his fifteen- and twenty-centimetre scissors, tapered and rounded; forceps long and short; knives, flat-bladed, serrated and hooked; shears, chisels and handsaws, manual and electric. Each one gleaming.

✵

There are two common ways to dissect a body. Karl von Rokitansky, a Viennese pathologist whose sideburns stretched from cheek to jowl, opened up more than thirty thousand cadavers himself to classify in words and pictures the origins of disease and unveil God's secrets. He published in 1842 the first volume of *A Manual of Pathological Anatomy*, in which he described his technique of severing veins and arteries to quickly remove all of a man's vital organs in a single block.

"Sometimes I'd do the Rokitansky, depending on how much speed was necessary," Harvey said, "because it's just faster to take all the organs out in one go."

But that morning, even if bodies had been backed up in the morgue, there was no hurry, no need to rush the dissection of a genius. So Harvey chose the Virchow, the meticulous organ-by-organ method unchanged in more than one hundred years.

Rudolph Virchow was a star of nineteenth-century science who recorded one of the earliest cases of leukemia in 1845. In 1856, he became director of Berlin University's Institute of Pathology and his work there altered the course of medicine, identifying malfunctions in the human cell as the biological root of illness. He launched a journal of pathology, described the various types of cells and introduced the concept that each cell derived from another, inspiring whole new fields of research. His work appealed to a medical community eager to believe that modern science could understand and explain nature, and in 1858 Virchow authored the text *Cellular Pathology*, which documented his microscopic discoveries, from the fibres of malignant tumours to the chemistry of pus. All of it flowed from what he viewed in the corpses he dismantled.

Far from the ruckus emerging in the lobby, in the morgue as quiet as a chapel, Harvey clutched the scalpel and traced the history of Virchow. Tilting Einstein's head left, he jabbed first behind the ear, pulling his blade down the side of the neck, over the clavicle and curving inward beneath the collarbone. He repeated the incision from behind the opposite ear, and at the juncture, where the two slits met in the middle like a necklace above the sternum, he drew the scalpel down again, across the chest, across the abdomen, pasty, weathered skin giving like cold butter to the groin, until he had carved a Y shape into the torso of the corpse. Blood from the aortic rupture frothed in the cavity below—telltale evidence in the dead, in whom there is no more blood pressure, nothing but gravity to direct the blood's flow. "You could tell right away it had burst," Harvey said.

He turned the flaps of flesh and muscle outward over the chest wall, tugging gently and slicing connective tissue on the underside to peel them back so he could manipulate the sternum. With shears, he bore down on the costal cartilage, calcified with age, that connected each rib to the breastbone, so he could hoist the sternum back and expose the organs beneath it. In keeping with the Virchow technique, the heart is generally the first organ to be extracted. He cut the drying arteries and veins and with his right hand palpated the heart to check for abnormalities before snipping it from its home, pulling the ripped aorta, from the chest to the abdomen, out with it. Atop the wooden table, Harvey turned the aorta over, analyzing the tear. No operation could have fixed that, he thought to himself. Cholesterol plaques coated the inner walls of the main artery. "His aorta was just riddled with plaques, the wall was just pretty much replaced with cholesterol, in a pale yellow," Harvey said. "That's what shortened his life, a fatty diet."

Harvey pumped the peritoneal cavity and measured 3.5 litres of blood. The aneurysm's flood had submerged the other organs, which would have produced the symptoms of a gallbladder attack. "This was definitely the result of the rupture," Harvey said.

He turned to the lungs next, cutting their attachments to the clavicle, and laid them on the table. He sliced through them laterally with a flat blade, so that each air sac opened like a book. Black patches coated them, but the interior still looked pink. He was lucky for a smoker, Harvey thought. He studied the neck organs in their place and then moved down the table to remove the stomach, kidneys and intestines. He fondled the liver, jaundiced too from years of overwork, then he scored it with parallel and vertical cuts to examine its interior. He weighed the tiny adrenal glands, fingered the pelvic organs and detected morphine in the bile of the green sac that was the gallbladder.

In Harvey's mind, no question lingered about what he should do next. Both Virchow and Rokitansky call for the removal of the brain at the final stage of the autopsy, so it can be weighed and described in case brown splotches stain its surface, indicating a stroke, or the lump of an undiscovered tumour hides in its folds. Harvey had removed more than sixty brains from autopsied cadavers in the previous year. He would later recall that Otto Nathan, no doubt battling grief and the numbing shock bound to accompany the spectacle of watching a close friend dissected, stood wordless as he placed a steel bucket below Einstein's listless head. Nathan had a reputation for being quick to temper, but if he had any objections at that solemn moment he never voiced them. Only later, when the impression lingered that Einstein's executor knew nothing of the brain's removal, did Harvey feel compelled to explain that "Otto Nathan saw everything I was doing. He watched me take out the brain."

To Harvey, taking brains was routine. In this case, in particular, Harvey thought it would have been shortsighted to leave it. "To me it was obvious that the brain of this man should be studied. Here was the brain of a genius—I thought, 'I better do a good job.'"

Einstein's dandelion locks dangled in limp white strings from the table's edge, and between the clumps plastered behind the earlobes Harvey combed an arch-shaped parting across his celebrated head. Not knowing that Einstein's body was to be cremated, he combed the hair carefully to avoid cutting it, which would have made the funeral home's reconstruction job more difficult. Harvey pierced the scalp at the point of his first incision and followed the parting with his blade, blood trickling into the bucket below. Then he pulled half of the severed scalp backward toward the base of the neck and the other half forward, so that a mat of hair smothered Einstein's face. The dark flesh of the temporalis muscles that anchor the scalp sat in circles on either side of the skull, parallel to the forehead. Harvey cut them from their margins to hike those forward as well, revealing the ivory vault of the skull.

A manual saw allows the tactile privilege of determining bone consistency, but the sawing itself demands strong, sure hands, steady enough to persevere through the time it takes to crack the skull, steady enough not to apply so much pressure that you slice through the precious matter beneath it. Electric saws, albeit noisy and dangerous for the spray of bone dust they raise, are quicker and easy. Harvey plugged his in. Metal teeth whirred through the cranium at the line where the brim of a crumpled fedora had once sat. It hummed relentlessly for several minutes until the blade burst through. Then Harvey repositioned it to cut a V into the back of the skull, slashing upwards to meet the first incision. He picked up the chisel then, wedging it into the groove his saw had created. He hammered it hard with a mallet and rocked it carefully, back and forth, between the halves of bone until the skull gave and he could pry it open. With patience and force it came off in the shape of an uneven triangle, and Harvey could see the tough, fibrous membrane of the dura, the outer-most of the three covering layers of the brain.

He cut the dura and the layers below it in places where his blade was least likely to contact the organ it protected, and the sheathing fell away. He tilted Einstein's head from side to side to gain more purchase as he gripped tweezers to pull the dura back over the brain toward the shoulders. He saw it then: a shimmering pearl in its cra-nial shell.

By its frontal lobes he pulled it gently backward, and then he snipped the tape-like strands of the olfactory nerves. He severed eleven other pairs of cranial nerves that wired the eyes and arteries to the brain. Behind the bridge of Einstein's nose, he inspected the fragile stalk of the cherry-shaped pituitary gland. Though Descartes believed it to be the seat of the soul, scientists had discovered that it served the hypothalamus. This one would have powered the physicist's growth, appetite and legendary lust with hormonal bursts that also flushed him with the adrenaline he needed to forge on with his figures.

Harvey drained the cerebrospinal fluid with a blunt-nosed pipe. Then he reached past the ropy stem of the medulla oblongata at the hindmost part of the brain and down into the spinal column, stretch-ing his fingers as far as they could extend—"about two inches"—and cut the fibres that wed the brain to the rest of the body. Then, with two hands, Harvey freed it from the skull and gingerly lifted it out.

"It looked like any other brain," he said.

After it is soaked and pickled, a brain can be as firm as a Christmas cake. But the fresh adult brain is gelatinous and slippery, covered in mucus and a web of veins. Harvey knew there was no point in exam-ining it then. He cradled it above his own head and placed it on the grocer's scale. The needle flickered to 1.230 kilograms (2.7 pounds), a bantamweight for such a mental giant, lighter than the brain of German philosopher Immanuel Kant but heavier than the one belonging to the founder of phrenology, Franz Joseph Gall. Carefully, Harvey carried it to the deep sink beside the high utility shelf.

"Because it was Einstein's brain I embalmed it very carefully, so it would be in a good state to start with," he said. "I ran the embalming fluid into the brain via the arterial system. Ordinarily I wouldn't be doing that . . . I would just sit the brain in a jar of formaldehyde. I had a container of formaldehyde up high and I was just sort of using gravity to force it through."

He connected the hose to the internal carotid artery that had run up from the spinal column to feed the front portions of the brain, the eyes, the forehead and nose with its curling branches. The warm pickling fluid seeped into the organ, gushing through the vessels and cooking many of its proteins and chemicals like heat to the white of an egg.

He hooked a string to the brain's base afterward so that it could dangle upside down in a large jar of formalin, a gentler formaldehyde solution diluted with water and saline. Had it not been suspended, the brain would have flattened under its own weight. Like a watermelon, roughly 80 percent of it was nothing more than water.

✵

By mid morning, cars full of newspapermen and a CBS television truck waited outside Einstein's house. They filmed people coming and going and canvassed neighbours for interviews. Few obliged, knowing how Einstein had hated publicity. More reporters turned up at the hospital as the day wore on, fraying the nerves of medical staff with their persistent questions. Kauffman told them that only the pathologist could confirm the cause of death, which would not be known until the autopsy was complete. Then he marched to the morgue to hurry the man of the hour.

Harvey knew he would have to speak to the press. But he thought the undertakers from the Mather Funeral Home were waiting to prepare Einstein's body for an open casket, so first he randomly placed the dissected organs back in the open torso to keep it from sagging and filled Einstein's empty skull with cotton wool batting. He replaced the severed cap of the cranium and stitched the scalp together back over it. The brain he left marinating in the back room of the tissue laboratory. Then he washed up, donned his white lab coat and went to meet the media crowd. On his way out he told Kauffman that he had removed the brain for study. His boss was pleased. Kauffman thought it would look good for Princeton Hospital to be examining the brain of Albert Einstein; it would help put the institution on the map.

☼

Upstairs, Einstein's friend and ophthalmologist Henry Abrams was unaware that the world had lost its prized thinker and he his treasured companion. He had been seeing patients all morning in the office tower when a frazzled nurse popped in.

"Einstein died in the middle of the night," she told him. "An autopsy has already been done."

Abrams ran from his office, down the stairs in shock. He could not bear to miss the opportunity to say goodbye. Attending a memorial service would have been enough to mark the passing of any other friend, but not Albert Einstein, the genius oracle to whom the public turned, to whom he turned for guidance and loved so well. Einstein had feared how strangers would mourn him, yet now it was a friend who scrambled through the corridors to worship at his bones.

"I rushed over to the hospital to see what was going on, just like a lot of other doctors did. The place was crazy. People everywhere. Everyone seemed to be in on the news, everything seemed to focus on his death."

He spotted Helen Dukas and Einstein's stepdaughter Margot standing in the hallway, visibly upset. Otto Nathan was there too. "They were saying that in Einstein's will he wanted to be cremated and that there should have been no autopsy." Abrams expressed his condolences but then continued toward the morgue, where he joined the others.

☼

Through many eras, and in many faiths, power is ascribed to the body parts of the especially pious. The ancient Greeks, the followers of Buddha, the Zoroastrians have all believed in their sanctity. According to eighth-century Catholic doctrine, relics themselves were not to be adored but honoured, in the belief that by respecting a saint's body "many benefits are bestowed by God on men." For many centuries corpses have been routinely sectioned and body fragments distributed for display in churches and cathedrals. As late as 1951, church officials brought the bones of the English Saint Simon Stock, who died in France in 1265, back to his homeland for veneration. Religious leaders used official Latin phrases to distinguish relics from the bones (*ex ossibus*), from the hair (*ex capillus*), from the stomach (*ex praecordis*), from the flesh (*ex carne*). So widespread did the practice become that collecting a complete set of any saint's mortal remains became something of a rarity. Catholics in Normandy clamoured for a piece of Mary Magdalene's preserved arm. When Saint Elizabeth of Hungary died in 1231, mourners paying their final respects lopped flesh from her corpse as she lay in state. The more popular the saint, the more people wanted a part. Swindlers sprang up to supply the superstitious masses with costly fakes.

In time, the line between piety and celebrity blurred so that relics were gathered not just from the religious but from anyone of renown. An American urologist allegedly keeps Napoleon's penis. The Chapelle des Ursulines in Quebec still displays the skull of General Louis-Joseph de Montcalm. The middle finger of Galileo's right hand, detached from his corpse by a souvenir-seeker in 1642, rests in a Florence museum. And if science was the religion of the twentieth century, binding people together with its own dogma and rituals, preaching the power of theories instead of prayer, then its secular messiah lay in Harvey's green-tiled sanctuary. Science had nibbled at the foundations of religion as soon as it began to explain the mysteries of nature, and Einstein had chewed off one of the biggest chunks. He was the accidental and eminently quotable prophet, uniting the world's war-weary under a cosmic banner that transcended the volatility of national politics. He preached equality. He championed the causes of underdogs, lobbied for the wrongly convicted and himself suffered Nazi persecution. An admirer once called him the greatest Jew since Jesus. Einstein himself joked that he had become a Jewish saint.

Several doctors gathered that day in their white coats, encircling Einstein's corpse, to see, to touch, to take. Most were strangers to him, come like pilgrims to pay homage, giddy even to be in his lifeless presence.

"Other doctors came in to pay their respects, I suppose, or just to see what was going on. They were all in there. They couldn't help themselves," Abrams said.

He saw the stitching across Einstein's scalp, and Kauffman told him that Harvey had been given permission to remove the brain. Abrams did not see Harvey, and he thought that he might have been in one of the back rooms. "I said to Jack [Kauffman], 'The eyes are considered part of the brain, you know,' and I asked Kauffman if I could have permission to take them. And he said, 'Absolutely, Henry, you have permission to take them.'" Kauffman, who never told his version of these events, perhaps felt grateful to the young ophthalmologist for bringing Einstein as a patient to Princeton Hospital in the first place.

Abrams paused for a moment over Einstein's face, gazing at his son's godfather, at his dear friend. Then with bare hands he lifted the eyelids to expose the dark, vacant pupils at the centre. The ones about which the French newspaper *L'Humanité* had raved in 1922: "Oh, those eyes! Those who have seen them will never forget them. They have such depth! One might say that the habit of scrutinizing the secrets of the universe leaves indelible traces."

Abrams pressed his fingers into the sockets to force the eyeballs forward, tucked the blade of his scissors in behind to reach the optical nerves that connected them to the empty head and snipped carefully. With forceps, he plucked the orbs from their sockets and plopped them into a jar filled with formaldehyde. Even years later, he marvelled at the ease with which it was done. "It only took a few minutes," he said.

Abrams said he later asked Guy Dean for written verification that he was now in possession of Einstein's eyes, and the general practitioner provided him with a letter.

"Dr. Abrams," it was reported to have said, "was the ophthalmologist who enucleated the eyes when Dr. Tom Harvey removed the brain at the autopsy"—as though the floating globes would somehow find themselves on the dock at Sotheby's and require authentification.

Abrams is retired now and winters in Florida, but he keeps the eyeballs still, in a jar, in a safety deposit box at a New Jersey bank. "Whenever I look at them, in my own mind, I see the greatest person of our time who did so much for our civilization. I felt that he should not just evaporate and disappear like that. I felt like I wanted to continue to be close to him."

Abrams knows full well that the late genius, his friend, would not have approved of his corporeal worship. "Einstein's will said he didn't want anything removed and Einstein was of sound mind when he wrote that he wanted no such thing," Abrams admitted. "I never discussed anything about what happened in that autopsy room with anyone. We agreed we wouldn't discuss it."

Harassed by requests for his autograph, Einstein had once remarked that mere signature hunting was the last vestige of cannibalism: "People used to eat people, but now they seek symbolic pieces of them instead."

Abrams will not say who else was there that morning and he does not know if Kauffman granted permission for the removal of any other keepsakes. Rumours trickled out that Einstein's heart and intestines disappeared in a bucket, along with his official autopsy report. Tom Harvey has always maintained that he never saw anyone else in the morgue except Nathan. But for at least an hour, Harvey was not there. He was standing on the front steps of the hospital talking to reporters, contributing something to the annals of science for the first time in his life, squinting into the noonday sun.

FOUR

PROMISES, PROMISES

☼

THE DEAD PHYSICIST gazed out at the world from the front page of the *New York Times* the next morning. It was the image Benjamin Cortizano had painted and presented on his last birthday: Einstein looking somehow melancholy, his head bent slightly forward, spectacles perched on the end of his nose.

Tom Harvey's middle boy, Arthur, noticed the newspapers at the breakfast table and there, in black and white, he saw his dad. Harvey was wearing his lab coat and speaking into a microphone on the hospital's front stoop. A reporter stood beside him jotting down his words, while another brandished a tape recorder. Arthur Harvey thought his father "must have done something important, something newsworthy to be there." He asked his mother about it before trundling off to school, pleased and no doubt proud to have something to share with his fifth-grade class.

By day's end almost everyone knew that Tom Harvey had taken Einstein's brain. If they hadn't heard it on the streets, they would read about it in the newspaper—as Einstein's family did. Louise Sayen, who lived next door to the Einstein house, called it a disgrace that stunned the family. "They were horrified and didn't think the Hippocratic Oath included anything like that," she said. "There was a lot of high feeling about it in the town, that this should have been done. It was very distasteful." Princetonians wondered what body parts they might lose if they died at the local hospital. "As word got around," she said, "people thought, 'There but for the grace of God go I.'" That Princeton prided itself on its indifference to celebrity, even when it came to its most famous resident, made it all the worse. "It was an incredible thing for a town that had treated Einstein with a hands-off manner," Jamie Sayen said. "Then everyone finds out the astonishing thing that there's been this feeding frenzy."

Over the years, details of what happened in the hours after Einstein's post-mortem would become as convoluted as a human cortex. Various people would recall the events very differently. Yet it is Harvey's own recollection that casts the most unflattering light on his handling of events. He admits that he never spoke to the family at all the day Einstein died, though he intended to and later wished he had, if only to have avoided the turmoil that followed. As Harvey tells it, the family discovered that he took the brain only when they read it in the next day's paper, and only then because Harry Zimmerman, or perhaps someone at the hospital, leaked the news to a *New York Times* reporter.

Scrummed outside the hospital entrance, Harvey told reporters "a big blister on the aorta which broke finally like a worn-out inner tube" had caused Einstein's death. But that was all he told them. Then he went back inside to the morgue. Einstein was already gone. Employees from Ewing Crematorium had whisked the body away, and Harvey now had time to think. He knew he needed help. He had no delusions about his own expertise in brain analysis. If Frederick Lewy had still been alive he might have called him in Pennsylvania. But as it was, the only name that came to mind was Zimmerman's. He wasn't aware that Zimmerman and Einstein had known each other, or that they had discussed studying the physicist's brain. But Zimmerman had been the first to introduce him to brain studies and the tantalizing tales of Oskar Vogt. In his later years, Harvey would say that Vogt's study on Lenin's brain was what inspired him to believe that Einstein's was worth saving. So he telephoned New York.

☼

Even after becoming director of the Albert Einstein College of Medicine, Zimmerman continued to do lab work at the Montefiore Medical Center. He still enjoyed playing mentor, and he surrounded himself with students and assistants. One of these was the young Robert Terry, who likely reminded Zimmerman of himself. Hungry to keep pace with advances in neuroscience, Terry went to Paris in 1953, where he mastered electronmicroscopy. Developed by Germans in 1932, the electron microscope magnified ultra-thin tissue slices to several hundred thousand times their actual size, revealing fresh anatomical details, particularly in the cellular structure of the brain. But Zimmerman lured Terry back with a promise to buy him an electron microscope at Montefiore. The two of them were in the lab that afternoon when the phone rang. Zimmerman answered it, and Terry heard his tone grow more animated. Must be an intriguing call, he thought.

"That was Tom Harvey, a former student of mine from Yale," Zimmerman said after he hung up. "He told me Einstein died today at Princeton Hospital where he works and that he's removed his brain." Whatever grief Zimmerman felt upon hearing of the death of the man he had convinced to lend his name to the college seemed displaced by the prospect of studying his brain.

"Harvey apparently said that he knew Zimmerman was the neuropathology expert—that he himself was no neuropathologist—and that he would send the brain to Montefiore Hospital," Terry later recalled. "Harvey said he would send it right away, and I remember Zimmy got off the phone and told me, and we were very excited."

✷

The item appeared in the *Times* not as a sensational scoop or tale of modern-day grave robbery but as a simple fact, slipped in amid the reams of newsprint commemorating the life of the Nobel laureate: "The body was cremated without ceremony at 4:30 p.m. at Ewing Cemetery in Trenton after the removal, for scientific study, of vital organs, among them the brain that had worked out the theory of relativity and made possible the development of nuclear fission." None of the other "vital organs" were identified in the press reports of the day. But the family had presumed the entire body had been incinerated. Hans Albert telephoned Princeton Hospital to complain.

"He was very upset," Harvey admitted. Hans Albert told Harvey that his father had not specified any study of his remains and that he would not have liked the publicity. Harvey said he was heartily sorry for offending the family. But "Hans Albert had given permission for an autopsy," Harvey explained. To his mind, standard autopsy procedure included removal of the brain and, in some cases, keeping it. Hospitals frequently retained tissue and organs from the cadavers that passed across their dissecting tables, for teaching or their own studies. Princeton was no exception.

Nowhere has the ethical code around patient consent and scientific research been fuzzier than in the autopsy suite. From medieval dissections to nineteenth-century body snatching and into the biotechnology conundrums of the twenty-first century, the legal status of a corpse remains ambiguous. Since no one can be said to own a human body, generations of lawmakers have not considered it property. Body snatchers were prosecuted not for theft but for outraging public decency. Between the world wars, the U.S. military commandeered the corpses of servicemen killed in battle to train its surgeons without so much as a nod from their devastated families. By necessity, doctors are schooled to consider the body as an object, not a sacred or spiritual entity. How else to deal with the horror of sawing, slicing, cracking and gutting another human being? Harvey had a natural aptitude for suppressing emotion. He could divorce himself completely from the daily demands of what, to any layperson, was decidedly dirty work. Many thought him guilty of a ghoulish deed. But to Harvey, a human corpse was a medical tool by which to learn, and Einstein's brain was the precious raw material of a career-making scientific endeavour. "It was the brain of a genius," he once said. "I would have felt ashamed if I'd left it."

On the phone to Hans Albert that morning after, Harvey stressed the scientific value of keeping the brain. He hoped to have it examined for anatomical signs of intelligence, he said, to see how it might be different from other brains. Hans Albert worried about the publicity that would follow the project. But the Quaker Harvey made a promise then that convinced Hans Albert his father's brain would rest in dependable hands. "I told him I would take good care of it, that I would not exploit it or expose it to publicity." In an oath that would outlast every other he would take in his life, Harvey solemnly vowed to become the organ's conscientious guardian. "I promised it would only be used for scientific study and that reports about it would appear only in scientific journals."

✵

Hans Albert, a successful hydraulics engineer, understood well the detachment that science demands. It was one of the few things his father left him. According to biographers Roger Highfield and Paul Carter, Einstein taught Hans Albert to show a stoical resistance to sentiment. The year before he died, Einstein made the uncharacteristic gesture of writing his son a birthday letter: "It is a joy for me to have a son who has inherited the main traits of my personality," Einstein wrote, "the ability to rise above mere existence by sacrificing one's self through the years for an impersonal goal. This is the best, indeed the only thing through which we can make ourselves independent from personal fate and other human beings." In his will, Einstein certainly displayed no special sentiment toward his son. He left Helen Dukas, his housekeeper and assistant extraordinaire, $20,000, all his books and personal effects and, for as long as she lived, the net income his copyright fees and royalties would generate. To his stepdaughter Margot he also left $20,000. But to his hospitalized son Eduard, Einstein bequeathed $15,000. And Hans Albert, whom Einstein considered to have a good job at Berkeley, received only $10,000. If the elder son wanted a personal memento from his father's belongings, he would have to ask Helen Dukas.

Annoyed though he was that his father's brain had been taken without permission, Hans Albert agreed that it could be left to science, in the custody of Thomas Harvey. Otto Nathan felt otherwise. Jamie Sayen, who would later become close to the executor while writing the book *Einstein in America*, said, "Hans Albert gave the permission and Nathan felt he had to live with it." Before that April in 1955, Nathan and Hans Albert had not been well acquainted. But Nathan soon regarded Hans Albert as a member of Einstein's troublesome first family, one he would sooner forget. "There wasn't any good blood between Hans Albert and Otto Nathan," Thomas Bucky said. "Nathan was the czar of the Einstein legacy." Nathan felt it was Einstein's second family, including his stepdaughter Margot, Dukas and himself, who had the late physicist's best interests at heart. And he, as the will implied, was the patriarch. Sharing joint control of Einstein's entire literary estate, he and Dukas alone had the power to approve—or disapprove, as was often the case—the publication of any word from the physicist's personal papers. Not until their say-so would these rights pass, along with all financial proceeds, to the Hebrew University in Jerusalem, in accordance with Einstein's will. In the meantime, they pledged to preserve his memory unblemished. Future scholars and biographers would snidely refer to them as the "Einstein priests." But Jamie Sayen felt sympathetic to the difficult position that Nathan had been left in. "Einstein had entrusted him with an impossible task, to be the keeper of the flame after the flame flicked out. In a sense he would always be caught between what a dead man wanted and what the scholars and biographers demanded."

As tributes to the late Nobel laureate poured in from around the world, Otto Nathan ensured that no ink was spilled over the post-mortem controversy. "Einstein's Son Gave OK to Study Brain," said the *New Jersey World-Telegram and Sun*. "Son Asked Study of Einstein Brain," read the *New York Times* headline on Wednesday, April 20. Hans Albert himself was never quoted in the stories. Everything was filtered through Nathan, who had picked the casket, paid the funeral and cremation bills totalling $350 and presented himself to the press as the family's official spokesperson. "The decision to preserve Dr. Einstein's vital organs for study was made by a son, Hans Albert Einstein," Nathan told reporters, "who felt that would have been his father's wish." Nathan disguised whatever reluctance he might have had, pointing out that Einstein had remarked "from time to time on the usefulness of the human body after death."

By then Nathan had already decided that he, not Hans Albert, would police the agreement made with Thomas Harvey. He intended to make quite certain that the Princeton pathologist handled the brain in the most discreet, judicious manner, for he, too, had made a promise.

✻

Harvey had wandered, unsuspecting, into the first clashes between Einstein's son and his executor. But he had secured Hans Albert's consent. And Otto Nathan, as he said, had been there in the morgue, watching as he sawed through the skull: "He knew the brain was being removed, but I guess he thought I would just put it back in with the body."

Local opinion might have been against Harvey, but the press celebrated his rise from small-town pathologist to the doctor who would lead the study of the century's hallowed brain. Nobody then questioned his qualifications for the job. His association with Einsteinian grey matter thrust him into the spotlight, and anyone who had ever been associated with him wanted a share of it—just like a schoolboy boasting about his dad. The local newspaper in Meriden, Connecticut, where he had interned during his Yale years, informed its readers that the same Dr. Thomas S. Harvey who would take part in analyzing the brain of Albert Einstein "once worked here" as Meriden's pathologist. Farther west, in his mother's native Iowa, the *Ottumwa Daily Courier* toasted him as a favourite son: "Ottumwans' Nephew to Direct Einstein Study," it said. Dr. Harvey was the son of former resident Frances Stoltz, the paper reported, whose brother Oscar Stoltz, Dr. Harvey's uncle, was the mayor of Ottumwa. Fruits of glory blossomed on the hometown family tree. Elouise pasted all the articles into a thickening scrapbook.

The newly anointed keeper of Einstein's brain accepted his part with the enthusiasm of a student handed the lead role in a school play. "The study will be made by a team of outstanding medical men," Harvey told reporters. Media requests to interview him flooded Princeton Hospital, so many that he announced on April 21 that he would hold a press conference the following Monday afternoon. He never considered this a violation of his promise that there would be no publicity; he would be talking about the logistics of the study, not Einstein's brain itself. In the run-up to the scheduled event, reporters wanted to know who would be working with Harvey. Not being especially well connected in neurology circles, Harvey mentioned only Zimmerman. The press assumed there were others, but that Harvey had simply declined to make their names public.

☼

After the thrill had subsided at the Montefiore lab, Harry Zimmerman and Robert Terry sat wondering what they would do with Einstein's brain once they got it. As an extreme example of intelligence, they agreed it was definitely worth a look. They just weren't sure what they should be looking for, exactly. They were familiar with the anatomy of disease. They routinely diagnosed Alzheimer's disease in post-mortem brain tissue by the telltale tangled axons. They could suss out a victim of multiple sclerosis by the frayed patches they discovered in the myelin sheaths that coat nerve fibres. But what clues did they have about the biology of genius? Oskar Vogt claimed to have found evidence of it in Lenin's brain, but Zimmerman likely regarded the study of a single specimen as meaningless. Their best hope of finding anomalies in Einstein's brain would be to somehow compare it to others even if history worked against them.

Crude comparisons among the brains of the talented, the ordinary and those the early neuroscientists indelicately dubbed "idiots" and "morons" had odious roots. In the beginning it was all about size. While phrenologists pictured the brain as neat little parcels controlling different functions, on the other side of the fence were scientists preaching against the organ's partition. Convinced that the whole brain governed all abilities, they reasoned that a larger brain suggested a smarter owner. So away they went with their weigh scales, measuring tapes and sacks of millet seed, filling empty skulls to determine "cranial capacity." They took no account of other factors that can affect the size of a brain—like malnutrition, body size, disease or the natural shrinkage that follows the preserving of a waterlogged organ. But it hardly mattered. In many cases, these nineteenth-century white male scientists fudged figures to justify their own view of the social hierarchy: women, non-whites and criminals ranked at the bottom of the scale. Other researchers, meanwhile, were collecting troublesome data: the brains of hardened criminals were not necessarily smaller than those of good men. Common day-labourers often had brains just as large as vaunted professors. "Idiots" sometimes had the biggest brains of all. Instead of comparing overall brain size among racial and cultural groups, scientists then began focusing specifically on the anatomy of leading minds.

German pathologist Rudolph Wagner undertook the first known investigation at Göttingen University in 1860, after he was bequeathed the brain of genius Karl Friedrich Gauss. The astounding German mathematician, who died in his sleep at seventy-eight, was said to have corrected his father's accounting mistakes at the age of three and shocked his elementary school teachers by adding up the numbers between one and one hundred within seconds. Gauss produced the first fundamental theorem of algebra, laid the foundation of modern mathematics and taught himself Russian at the age of sixty-two. His development of non-Euclidean geometry later inspired Einstein to conclude that gravity itself must be fundamentally geometrical.

Wagner reported that though Gauss's brain weighed only slightly more than the average, at 1.492 kilograms, the intricate pattern of its wrinkled surface was richer than that of any brain previously recorded. Yet when Wagner studied five other brains belonging to eminent Göttingen professors, he found that some weighed considerably less than average and none had complex convolutions on its surface. Wagner concluded that neither size nor surface pattern was a reliable indicator of intellect. But many scientists didn't buy it. According to neuroscience historian Stanley Finger, Wagner's work prompted all sorts of academics to make arrangements for their postmortem brains to be studied "in a kind of surrealistic competition to see who had been endowed with the best brains during life." The contestants, mostly scientists and their colleagues, founded clubs, like the Mutual Autopsy Society of Paris and the American Anthropometric Society. But the results never yielded consistent links between intellect and brain anatomy, only debate. Nevertheless, in the early 1900s, the prominent anatomist Edward Anthony Spitzka, future president of the American Neurological Association, described the brains of more than 130 gifted people—including Beethoven, Bach and French philosopher René Descartes—still confident that size and convolutions held clues to genius.

Despite this spectacular scientific inertia, some of which the press recounted in 1955, no one suggested abandoning the study of Einstein's brain or incinerating it like the rest of him. "Moving picture cameras will record every step in the analysis. The most up-to-date methods will be used in an exhaustive search for clues to the centuries-old question of the origin of genius," one report gushed. "The brain that, in a living body, vastly extended the horizons of man's understanding of his universe, may, as a result of the studies, open new vistas of knowledge after its death," said another.

In the echo of the atomic boom, society had undergone a conversion. It expected miracles from laboratories. Faith in science had become as fashionable as Brylcreem and Hula Hoops after the Second World War, and all around were testaments to its endless possibilities: commercial jets crossing oceans and continents; colour televisions; ENIAC, the world's first computer, flickering on its 18,000 vacuum tubes at the University of Pennsylvania. Medical science held particular promise. Harvard surgeons had successfully transplanted the first kidney. North America was rolling up its sleeves for the molecular magic of the polio vaccine. Streptomycin was curing tuberculosis. Infections, the early century's number-one killers, seemed destined to disappear altogether; antibiotics could literally chew up bacteria and spit them out. In England, scientists James Watson and Francis Crick made their discovery of deoxyribonucleic acid, or DNA, the hallowed blueprint for human life. What's more, the structure of heredity came in a form the popular imagination could easily grasp—two spiralling ribbons of genetic code.

In brain studies, the story was similar. Since the 1929 discovery of acetylcholine, one of the chemicals that powers muscle, researchers had begun to investigate how electrical pulses prompt brain cells to produce certain chemicals. With cutting-edge tools like the electron microscope, they quickly understood that communication among neurons is predominantly chemical, not electrical. They reasoned that serious mental diseases were the result of chemical disruptions in the brain, and the focus of brain science began to shrink from the 1.3 kilogram mass, to the microscopic cell, to the molecule. If chemicals powered the brain and behaviour, what, researchers scrambled to find out, could changing them do?

In Canada, world-famous neurosurgeon Wilder Penfield had seemingly gone inside the human mind, finding the places where memories are stored in the cortex of the brain. Operating on patients suffering from epilepsy at the Montreal Neurological Institute, Penfield had electrically stimulated the open brains of awake, though locally anaesthetized, patients. Jolting certain areas, to determine which regions to avoid cutting during surgery, Penfield made an accidental discovery. Zapping one spot triggered a patient's recollection of a favorite melody. Zapping another prompted a vivid reminiscence of the view from a childhood window. In the same way, he could spur toes to wiggle or fingers to twitch. He went on to chart a new map of the cortex, bolstering the notion that Broca, Vogt and indeed the discredited phrenologists had promoted: certain parts of the brain control certain functions and behaviours. Penfield himself did not believe that the mind was entirely a function of the brain. But if researchers could reduce things as esoteric as memory to a geographic location, people couldn't help but wonder if it might be only a matter of time before they unlocked the information inside another person's head. Was it so far-fetched to think that

researchers would soon solve the divine riddle of human intelligence? A kind of breathless confidence cloaked post-war brain scientists, as it had a century before. One report of the day suggested
that the feat of finding the source of intellect could take no more
than fifty to one hundred years. Half a century later, scientists would
still be stumped by the electrochemical symphony of thought. But,
as Einstein noted and Thomas Harvey would discover, the passage
of time is relative.

☼

Though Zimmerman was sensitive to Einstein's aversion to media
attention, it soon became obvious that he would have to tell the
media that Montefiore would soon have the great man's brain. Once
Harvey had mentioned his name to the press, reporters began calling the hospital for details. Montefiore officials told them that the
organ was due to arrive at 4:00 p.m. on April 19 for dissection and
analysis, although the exact source of this information was never
reported. The newspapers sent a photographer to take Zimmerman's
picture that day. He sat at his desk wearing thick black glasses, a lab
coat stretched across his beefy shoulders, smiling broadly over a
microscope. A portrait of Albert Einstein hung behind him on the
wall. "Dr. Harry Zimmerman to begin study of Einstein brain," the
cutline said. But two days later, the brain had still not arrived in New
York. Reporters called Jack Kauffman.

"The brain will not leave Princeton Hospital," Kauffman told them.
As for the expectation that Montefiore scientists would conduct the
actual study, he said, "There's just nothing to it." He told them that
the brain would harden in solution for a few days and "pathologists"
there would work on it.

Still strategizing to improve the hospital's stature, Kauffman likely
recognized the benefits the brain could bring to his institution. It

would undoubtedly inflate Princeton Hospital's reputation as a prestigious centre of scientific inquiry. The university and the Institute
for Advanced Study might hold Einstein's papers, but Princeton
Hospital would have his brain. Let it go, just like that, to New York?
The question itself must have seemed ridiculous.

But a spokesman at Montefiore insisted that they had been told
the brain was on its way, going so far as to point out that its study
would take into account the failings of similar investigations in the
past. "Weight, size, color, or complexity of the brain are not in themselves any evidence of super-intelligence," the spokesman said.
"Medical history is replete with evidence of geniuses who had small
brains and idiots who had giant brains."

Zimmerman called Harvey to ask about the delay.

"Harvey told him that the Princeton Hospital board refused to let
it leave, that there was no way they would part with such an important specimen," Terry said.

Zimmerman heaved a disappointed sigh and put the phone down,
grumbling something disparaging about meddling hospital trustees.

Harvey tried to downplay the controversy. He described it as "a
misunderstanding" when reporters called. He planned to meet with
Zimmerman in a few days to straighten it all out, he told them. It
was just a miscommunication, that was all. But he was sore at
Zimmerman for talking to the press, not that he had ever explained
to his former professor the promise he'd made to keep the brain
study under wraps.

If Harvey had truly intended to turn the brain over to
Zimmerman, the point was now moot. He would have caught hell
from his boss. Kauffman had been unhappy enough at the mere
suggestion of it. Years later, Harvey would maintain that all he had
wanted was Zimmerman's help—a learned peek at the pickled
specimen once he had completed the long ritual of preservation.
Then, he thought, they would be able to publish the findings in one

of the many medical journals he read and admired. Fate, as he saw it, had presented him with the professional opportunity of a lifetime. Of course he wasn't going to give it away. "It's not every day you get a genius to autopsy," he said.

Like everything else, the New York–Princeton turf squabble over Einstein's brain found its way into print. "Brains of Great Scientist Cause Hospital Dispute," read the headline over an Associated Press story that reported, "Pathologists couldn't decide where or when they would start their work." In the *Chicago Daily Tribune*, the wording was blunter: "Doctors Row over Brain of Dr. Einstein." Each side wanted to claim the brain for itself. The tug-of-war stories played out for the rest of the week, while contractors painted Einstein's little yellow house white to make it trickier for tourists to identify which one along Mercer Street had been his, accommodating the late scientist's wish that his former home never become a shrine. At the same time, the truth about the brain's removal was already blurring. Some reports mentioned that Einstein had actually willed the prized organ to science.

Harvey winced at the coverage. Not only did it tarnish his debut performance as an important man of science but it betrayed the family's zeal for privacy, which Otto Nathan had recently made abundantly clear. When Nathan read that Harvey had called a press conference for April 25 to discuss the brain study, Einstein's executor blasted him over the telephone. Cancel it, he told Harvey; a press conference was out of the question. Nathan hammered the fact that nothing about Einstein's brain was to be publicized—nothing. And now here was this mess. Nathan was sure to erupt.

The morning of the cancelled press conference, Nathan telephoned from New York again. He was livid. Harvey apologized and made every effort to calm him. He told him that he would go to see Zimmerman at Montefiore and make sure that the Bronx hospital clearly understood the family's desire for privacy. Nathan heard him out and said that he would call Zimmerman himself, but patience was not one of his gifts. On April 26, 1955, Nathan wrote to Harvey from his Manhattan apartment.

Dear Dr. Harvey:

I would like to confirm the substance of my telephone message to you of yesterday morning. . . . It is my understanding that a limited and conditional authority was given to your hospital and one or more of the physicians on your staff to make a suitable medical examination of portions of the brain (which had theretofore been withheld without authority), such authorization having come not from me nor by authorization of the decedent, but solely on the authority of Albert Einstein, Jr., one of the children, and that such authority was given upon the distinct understanding and condition that the matter would be effected in the strictest confidence and privacy; that no publicity or notoriety would be allowed to intervene; and that the results of the examination would be made available only in normal medical channels for the benefit of mankind.

Instead there has raged unseemly controversy between you and your hospital on the one hand, and Dr. Zimmerman and the Montefiore Hospital on the other, with you insisting that the breach has occurred through publicity issued by Dr. Zimmerman or his hospital, and the latter informing me conversely that they have issued no announcement or statement whatever.

As late as yesterday morning when I telephoned you you informed me that you agreed completely with my objections on the subject of notoriety; that you were to meet with Dr. Zimmerman at about 3:30 in the afternoon to discuss the matter, and would among other things call my protest to his attention, and that you concurred in my suggestion that I telephone Dr. Zimmerman at 4 o'clock to confirm my position. When I telephoned Dr. Zimmerman at that time, he informed me that he had no appointment with you, that you had not arrived as yet, and that he did not even know whether to expect you. Later I was advised by him that you had come to see him.

I am at a loss to understand the various discrepancies and feel completely insecure about the various undertakings and promises made, and all of us are in addition gravely disturbed at the utterly distasteful notoriety, which would have shocked the late Professor Einstein beyond words.

The permission and authority heretofore granted you or your hospital or any one connected with it is accordingly herewith suspended and demand is made that such remains as were retained be not further used or employed in any way in connection with any examination, analysis or study of any kind, and that the same be carefully set aside for disposition according to further instructions which you will receive.

I assure you, Sir, of great regret that this step becomes [necessary] as it seems to us necessary under the circumstances, and in order that the wishes of the deceased and family be properly respected.

Yours very truly,
Otto Nathan

Nathan sent the letter by registered mail, along with a copy to Jack Kauffman. It must have left Harvey cold when he read it, because the next thing he did was completely contrary to his character. Confrontations were not his style. Yet he got in his station wagon, drove an hour from Princeton to New York and buzzed at Nathan's Tenth Street apartment, a block from Washington Square. "I went to pacify him," Harvey said. It was an old building of crown mouldings and soaring ceilings. Stacks of papers and books were piled high on the floor and on the shelves. Nathan often kept his blinds pulled down low in the study where he worked.

The last time Harvey had seen him, Einstein's executor had been on the pathologist's turf, the silent observer in his morgue, watching Harvey handle the mortal remains of Nathan's great friend. Now, Harvey was in Nathan's domain, with his professional ambitions in Nathan's hands. Harvey's recollections of the details of their discussion have grown hazy with time, but whatever he said, he managed somehow to appease the gruff-tempered executor.

The two men certainly had a good deal in common. They were both modest-living men who crusaded for social justice in their own ways —Harvey through the Rotary Club and the Individual Liberties Association, and Nathan through various political groups, published essays and the courts. Joseph McCarthy and the House Committee on Un-American Activities had targeted Nathan as a Communist in 1952, and the State Department had refused to issue him a passport when he considered finding professional work overseas after the war. Against his instincts, Nathan eventually swore an affidavit that he had never been a Communist. But the State Department would not budge. Nathan beat the feds in two rounds of court hearings, and he was at that point awaiting an appeal. Harvey knew of Nathan's troubles and told him that he sympathized with his cause. Both men were also devout pacifists, Nathan from experience, Harvey as a matter of faith. Each would likely have lived an ordinary life of anonymity, except that by some chance of space and time his destiny crossed the path of an extraordinary man. And both men, if Harvey had his way, would be entrusted with the profound task of preserving a part of him. "We were both kind of liberals," Harvey said. "I think I convinced him that I was the right person for the job."

There was something too about the study itself that appealed to Nathan. Harvey thought there was a chance that researchers would find Einstein's brain biologically different from those of other mortals. What better way, as the Lenin-crazed Soviets had felt, to justify Einstein's exalted legacy than to scientifically prove his genius?

People who knew Nathan, like Louise Sayen, said he would have supported a scientific investigation if he had thought it might bolster Einstein's reputation. "He absolutely worshipped the ground Einstein walked on," she said. Nathan was a passionate academic, so that if Harvey had played up his connections to the ivory tower, to Frederick Lewy and his years at Yale, it might also have helped sway him. Einstein historian Robert Schulmann, who would come to know Nathan in the following years, suggested that the executor had a vested interest in having science certify the late physicist as extraordinary. "There is a German word, *ehrharden*," Schulmann said. "It means to be raised above the pedestrian, the routine. If one is associated with the superhuman, the divine, it is *ehrharden*. . . . Nathan and Dukas thought Einstein was God's gift to the world and to be in this person's orbit was their great satisfaction.

"Nathan wanted to hear that Einstein was a Prime-A Genius."

The next time Otto Nathan wrote to Harvey, on September 22, 1955, his tone was markedly different. He wanted to know how the brain analysis was progressing, if there had been any results "negative or positive." But it was his closing remark that spoke volumes about the success of Harvey's mission to New York. "I hope that you had a pleasant summer," Nathan wrote, "and I look forward to seeing you one of these days in Princeton. Cordially yours, Otto Nathan."

※

Harvey realized that one wrong step could shatter the fragile bond he'd forged with Nathan and cost him Einstein's brain for good. Theirs was an eggshell friendship. So he never even considered popping the brain in the trunk the summer morning he drove to

Washington, D.C. If he took it, Harvey suspected it might be a one-way trip. The army had summoned him. Lieutenant Colonel Webb Haymaker, the head of neuropathology at the U.S. Armed Forces Institute of Pathology, had studied under Wilder Penfield at the Montreal Neurological Institute. He boasted a resumé that already included the examination of one notable specimen: the brain salvaged from the bashed skull of Benito Mussolini. Italy's anti-Fascists had gunned down their dictator in 1945, kicked in his head and strung up his corpse, feet first, alongside that of his mistress, in a Milan square. People had spat at him and poked him with sticks until the advancing U.S. troops cut his body down, took it to a hospital and extracted the debris of his brain. The army's neuropsychiatric unit sent it to Haymaker to search for signs of insanity, or genius, or syphilis—anything that might explain his tyrannical charisma. Haymaker found no evidence of any such thing, and Mussolini's brain would sit in military archives for twenty years before a U.S. diplomat returned fragments of it in a small box to his widow in 1966.

If Mussolini's brain represented a fiendish specimen that had fuelled war, then surely the brain of Albert Einstein, who had helped push the Allies to victory with a mere five-character mathematical equation, merited attention. Einstein might have been born German, and a little too left-leaning for a Federal Bureau of Investigation drunk on McCarthyism, but he was a national hero just the same. And Haymaker told the press that studying Einstein's brain would offer the chance to find "a general pattern for the brain of a genius." The enlarged size of a particular brain area, or richness of grey matter, or clear, unclogged arteries despite Einstein's advanced age might offer clues, he suggested. "It's in this sense that the study of Professor Einstein's brain is valuable," Haymaker said, "that is, it furnishes still another opportunity to try to add to the picture of what makes a great brain tick."

After all, the Soviets were seeking the same answers. The examination of talented brains in Moscow had become nothing less than a national secret after the war. The analysis of Lenin's brain filled fourteen leather-bound volumes in the institute that Oskar Vogt set up. They were locked tightly in guarded glass cabinets like classified documents. The brain of Josef Stalin, who had sanctioned the institute and the cloak behind which it operated, also ended up dissected and sandwiched between microscopic slides. Though Moscow had sent Vogt packing by 1930, concerned that his general theories on people's inborn differences clashed with Communist ideology, the Soviets continued to collect brains. Institute officials made plans to harvest the brains of gifted authors, composers, philosophers and scientists after their deaths. Under close political supervision, researchers probed for biological clues to personal power, achievement and genius. But whatever Soviet scientists gleaned from their efforts, which would turn out to be very little, remained hidden behind the Iron Curtain of the Cold War. Institute scientists were forbidden to publish. In 1967, 1969 and even as late as 1980, the Communist Party rebuffed letters from North American researchers interested in their findings.

Haymaker had called a meeting of top neuroanatomists to discuss how the study of Einstein's brain should proceed. Naturally, he extended an invitation to the man who happened to possess it. He fully expected Harvey to turn the organ over to Uncle Sam. Haymaker's eldest daughter recalled that her father was genuinely excited about taking a look at it.

Harvey scanned the faces of the scientists when he arrived. "Only two of them looked friendly," he thought. There were a half dozen of them sitting around a large table, sheaves of paper before them, pens in their pockets. At one end sat Haymaker, a lanky fifty-three-year-old, wavy hair cropped to regulation length, all arms and legs and long fingers fluttering over notes and an overflowing ashtray. Haymaker once confessed to smoking fourteen packs a day, but in truth he was more of a lighter than a smoker, butting out after just a couple of hauls so that ashtrays looked miniature beneath his stack of prematurely stubbed cigarettes.

Haymaker opened the meeting by introducing the eminent and eager researchers who had gathered: Gerhardt von Bonin, a skinny, balding, world-renowned German neurologist who had authored several books on the cerebral cortex; the eminent Jerzy Rose of Johns Hopkins University; Hartwig Kuhlenbeck, a specialist in invertebrate brains from Philadelphia; and Walle Nauta, an Indonesian professor from the University of Maryland who had just developed a silver stain that allowed researchers to trace degenerating nerve fibres in the brain.

The meeting began with the researchers discussing fairly standard proposals, Harvey thought. "They just wanted to section it and study it," he said. "One fellow suggested making a cast of it, you know, so you'd always have a three-dimensional image of it for measurements."

Harvey didn't say much in response.

So, they asked him, would he show it to them?

"I didn't bring it with me," Harvey replied. "The preservation is still in progress."

When would he send it over? they wanted to know, narrowing their eyes.

Harvey, likely sprinkling his sentences with his characteristic chuckles, told them that he intended to make the initial observations himself, take measurements and then dissect it. The truth was, he had no intention of giving it to them, ever. He wanted to supervise the job himself. But more than that, he imagined the reaction Nathan—a devout pacifist at war with the American government—would have had if he'd turned Einstein's brain over to the U.S. Army. "He sure would have exploded," Harvey said.

The members of the esteemed Washington gathering fixed a stony glare on the Princeton pathologist. "I was in bad graces with those fellas," Harvey said. "They wanted that brain and I wouldn't give it to them. It was all hopeful on their part that they would get it. They never tried any coercion to force me to turn it over, and I left."

But before he walked out, Harvey made a mental note of the friendly faces: Kuhlenbeck and Nauta. Those gentlemen would be worth tracking down some other time, he thought.

☼

Brains were all Harvey could think about in the weeks afterwards. How best to store and study them. How to cheat the vandalism of time. To bolster his limited expertise in the area, he sought guidance from the scant collection of reference books available, determined now to go it alone. Harvey himself had no burning desire to unveil a physiological explanation for Einstein's brilliance. He was not consumed by the need to know what had enabled a twenty-six-year-old patent clerk to scrawl the theory of relativity. Harvey's ambitions, which destiny had thrust before him on a metal slab, were not so lofty. He craved simply to make a notable contribution to a worthwhile scientific endeavour. It was enough to be "the supervisor" of the studies of Einstein's brain, to be remembered and respected for providing posterity with a well-kept specimen.

He took the train to Philadelphia and roamed through the medical bookstore. But there was precious little. The best brain work to date had been compiled in Germany and had been effectively lost to the world since the war. Virtually none of it had been translated.

Harvey settled on *Neurophysiology*, a 1928 text by John Farquhar, a Yale doctor who had studied under the great brain surgeon Harvey Cushing. Too bad Cushing himself had passed on, Harvey thought; he might have been the man for the job. But it was *The Isocortex of Man* that proved to be Harvey's bible of a brain book, his holy grail and the inspiration for his elaborate plan. The softcover tome, as long as his forearm, had been published four years earlier, in 1951, as the standard monograph on the adult human brain. Percival Bailey and Gerhardt von Bonin, one of the men Harvey had thoroughly disappointed at the Washington meeting, were its authors, world-renowned and respected neurologists at the University of Illinois in Chicago. Their text recorded a full description of the brain of an eighteen-year-old Indiana boy killed in a car crash. His body had crumpled, but his brain had survived to represent the average human organ for generations of scientists to come, the boy having left the world too young to prove his intelligence to be anything beyond ordinary. The textbook included diagrams for sectioning a brain, instructions for staining the organ's different elements and a healthy dose of skepticism about previous "cerebral mapmakers."

Alone, in the back lab at Princeton Hospital on nights and week-ends, after he had cut and cleared the steady stream of the hospital's cadavers, Harvey thumbed *The Isocortex of Man* and read the prophetic opening words of the first chapter:

> It must, however, be owned that all that both Ancients and Moderns have told us about the brain is so uncertain that their books which contain the Anatomy of this organ may be said to be chiefly a Collection of Doubts, Disputes and Controversies— but still a great advantage may be made of their Labor and even of their Mistakes.

He fished Einstein's brain from the bucket where it hung upside down, as resilient now as a ball of mozzarella. He weighed it once again in the grocer's scale suspended from the ceiling: it had already shrunk by 135 grams. He placed it on the short wooden table that sat atop the stainless steel of the autopsy bed and plugged in a stage-light lamp to illuminate it. He loaded film into his Exakta 35-millimetre camera and snapped and clicked, then circled it to snap again, cap-turing a profile from all sides. Like a cartographer, he sketched a freehand likeness of the brain and all of its neurological continents, measuring its landscape with calipers and jotting down the figures: its overall height (just under 9 centimetres); its width (7.5); and the expanse of the frontal, temporal and parietal lobes. In all, Harvey drew it from seven different vantage points, three from the right side, three from the left, and one from the top. The photographs and measurements would one day be precious, he thought, because he knew what he would do next. State-of-the-art brain studies of the day demanded slivers of an organ for microscopic studies. Harvey had to turn once again to his knives. But something stopped him. A stab of sentiment, perhaps, recognizing that these would be the final moments of it whole, the last instance in time of the organ's entirety. It seemed woefully inadequate then that he should be armed only with photographs and pencil drawings of what it looked like. They were for science, and Harvey seemed to want something entirely more romantic for himself, an enduring souvenir at which he could marvel through the years, recalling the first heady days when it was new and shimmered with promise.

A few years earlier, he and Elouise had hired a local artist to paint portraits of each of their three sons. It struck Harvey now that she could do a similar job with Einstein's brain as her subject. Anne Bonine Brower was so intrigued by the request that she said she would do it for free. She came in the evenings when the lab was quiet. Harvey propped the brain back up on the wooden table and she positioned her chair at a particular angle in his morgue and set to work drawing and shading on her sketchpad in black ink. Einstein might not have been entirely surprised to know that this posthumous body part had been deemed worthy of a still life. So often had sculptors, painters and cameramen banged at his door that the ever-ironic physicist once introduced himself to a stranger on a train as an artists' model.

It was a perfect likeness, Harvey thought when Brower had finished. "It's really beautiful," he said, though he never hung it, or had it framed. He just tucked it into his files with the photographs and reports.

Because he was still young then, Harvey never doubted that he would be able to manage a meticulous dissection without a machine. Often, neurologists sectioned a brain with a microtome, inserting the organ into the type of rotating blade a butcher uses to slice cold cuts. Princeton's microtome was not equipped for the job, but Harvey made do. "In those days, my hands were steady," he said. "I cut it myself, most of it there in the lab. I kept track of the pieces by number, which I labelled on to the sketches I drew." He sliced it first along the corpus callosum, severing the left hemisphere from the right. He took photographs of its exposed centre from each side and then weighed each of its two halves: the left at 550 grams and the right at 545. The left hemisphere, at 17.2 centimetres, was slightly longer than the right, at 16.4. Then he began his slicing of the halves, as though he were cutting the crust off loaves of bread, carefully squaring off chunks of the cerebral cortex, trying to follow the pattern mapped out in *The Isocortex of Man*. To each piece of sectioned tissue he pasted a numbered white label that corresponded with the numbers he wrote on his sketches, showing where each piece had fit. Not that he could imagine then that someone someday would want to reassemble the puzzle, but so that anyone who studied a fragment of Einstein's brain would know where in the organ it came from. One chunk of chopped grey matter can look just like another.

He wrote the number 1 on the sketch of the right hemisphere, just behind the cerebellum. Number 2 followed behind it, then number 3 and so on until he had rounded the back of the brain. Pieces 12 through 18 covered the top rear patch, and just behind the right frontal lobe began the smaller blocks of the 30s. According to the *Isocortex* book, he should have had about one hundred pieces when he was complete. But because he had also been looking at brain maps charted by the turn-of-the-century neurologist Korbinian

Brodmann, "my method turned out to be a something of a hybrid," he said. "I ended up with more than two hundred pieces." In fact, he cut the two-hundredth piece by the time he had finished with the left hemisphere. Some parts he left whole: the cerebellum, the arterial coverings and a fist-sized piece of the cortext. He placed them in a mason jar filled with formaldehyde. "I wasn't at the time aware of the cognitive function of the cerebellum," he explained. "I thought of it as something like a flywheel . . . that works to smooth out our motor actions and it might be too that, partly, I was a little too lazy to do that." The artistry exhausted him. After all, as Harvey pointed out, "I had to do all of this in my spare time. I had to earn a living." Yet it wasn't the sort of task he could leave and return to another day. He might lose his place—or worse still, a piece of the genius.

FIVE

PIECES OF GENIUS

☼

THE PHOTOGRAPHS SLID wet and sticky from a chemical bath in the darkroom of a photographer who worked for Princeton University's physics department. Howard Schroeder had read about Harvey's possession of Einstein's brain and offered his services out of the blue. In all, he developed some two hundred photographs and hardly minded the effort. The wrinkled landscape of Einstein's brain was probably the most intriguing image to emerge from his tray since he'd shot enemy topography for bombers during the war. He pegged each picture and hung it to dry. Harvey had told Schroeder that he couldn't pay him for the job. But the Einstein specimen had its own currency; Schroeder never asked for compensation.

Washington had started handing out money to fund science projects after the war, and college and university researchers were eagerly taking up the bureaucratic chore of filling out applications for government grants to pursue particular lines of study. But at a small hospital like Princeton, removed from academia's culture and protocols, Harvey never considered the possibility that a federal agency might fund the mammoth task that lay ahead. "It never occurred to me to apply for a grant," he said. Besides, having alienated the U.S. Army, and knowing Nathan's cynical view of the government, Harvey decided to keep Einstein's brain well away from officialdom. Instead, every aspect of the study would remain under his tight control. It was the way Harvey wanted it and, to his mind, the way Nathan expected it to be. To get things done, he relied on personal contacts, and through the summer of 1955 he plumbed those deeply. "I had done all I could do at Princeton," he said. The pictures had been developed, the measurements recorded and the brain dissected. The next step demanded the resources and expertise of a high-end laboratory. He needed microscopic slides. There was no other way for scientists to examine Einstein's brain cells up close. So he phoned his old lab colleagues at the University of Pennsylvania. "A place," Harvey said, "I could trust."

Einstein's brain took its first trip out of state in the back of the Ford. Its putty-coloured pieces, some chopped to the size of a big toe and others as small as a penny, bobbed in pungent formalin all the way to Philadelphia. There were so many of them that Harvey had to divide the brain tissue blocks between two large glass containers, the kind people usually fill with cookies and display on kitchen counters. When he called ahead, Marta Keller, a German technician who had followed neuropathologist Frederick Lewy to the States, promised Harvey that she would handle the brain herself. She realized that the process could take months of work on top of her regular duties, but, Harvey said, "She didn't mind, it being Einstein's brain."

Keller told him that some of the tissue chunks were still too big to work with. Harvey trimmed them down himself at the Penn lab, carving the genius brain into a total of 240 blocks and adding more numbers to his corresponding brain sketches. Keller promised to phone when the slides were ready. Then Harvey drove back to Princeton, for one of the few stretches of his life without the brain by his side.

To preserve the brain, to capture it as a still life of cells, Keller needed patience and the sure hands of an artist. Otherwise she might have blown herself up in the process. European scientists had been dunking tissue specimens in a nitrocellulose concoction since the 1880s. The resulting "celloidin" hardened tougher than gelatin, strong enough to seal wood and lacquer floors and make ping-pong balls. Human tissue embedded in the plastic-like substance can be sliced so thin it is practically translucent. The thinner the slice, the better the view from the microscope. But celloidin arrives as a highly flammable powder, or in maleable popcorn-looking clumps, unable to withstand prolonged shock or friction. Soldiers in Europe and North America once used it to produce "gun cotton" for the battlefields. So scientists held their breath working with it. By the 1950s, many North American scientists preferred to embed tissue in the cheaper, and safer, paraffin wax. Some considered celloidin outdated, even old-fashioned. But Keller was schooled in the German tradition and celloidin-embedding, though a dangerous and laborious process that consumes months, was what she knew best. And so the organ that unleashed the atom's explosive secrets would itself be locked within a combustible preservative.

Keller soaked the 240 pieces of Einstein's brain in tap water overnight to wash away the fetor of formalin residue. Then she immersed them in alcohol and water, increasing the proportion of alcohol over several weeks to slowly dehydrate the pieces. Next she submerged them in ether alcohol for two days to better absorb the nitrocellulose. The compound itself was dissolved by Keller in ether alcohol, mixing three batches: the first as thin as milk to penetrate the brain's nooks and crannies, the second as thick as table cream and the third as dense as corn syrup to harden smooth and clear. In jars screwed tightly shut, the pieces marinated in the thinnest concoction first. Then Keller repeated the process with the thicker batch. Finally, into small dishes only slightly larger than each chunk of tissue, she poured the thickest batch of the nitrocellulose mixture and placed a piece of Einstein's brain into each one. After she let them harden in the airtight confines of the lab desiccator, she covered them with alcohol to loosen the edges and popped the embedded pieces from their dishes like muffins from a tin. Each morsel of tissue looked as though it were encased within a hard, solid square of lemon Jell-O. Preserved in this way, the uneven blocks of Einstein's brain need never swim in formaldehyde again. Alcohol diluted with water would be enough to keep the celloidin, and the tissue within it, from dehydrating.

Only from embedded pieces can microscopic slides be cut. Even pickled tissue would wither like butter under the force of a blade. Round and round the razor-edged wheel of the microtome turned, slicing 1,000 slivers, each .12 millimetres thick, from the 240 pieces of a single brain, its matter reduced to ever smaller and smaller components, like a study in quantum physics. Keller shaved slide samples from each region of the cortex. She used Franz Nissl's recipe to stain some violet-blue and used a cadmium brown for others, sandwiching them between tiny rectangles of glass for viewing under a microscope.

☼

On September 29, 1955, Harvey replied to Nathan's friendly inquiry about the progress of the brain's analysis. Nathan's victory in court against the State Department had recently been making headlines: the judge had ruled that the government either had to pony up a passport or allow Nathan to cross-examine the witnesses against him. Harvey began his letter by congratulating the executor of Einstein's estate on his victory. "I feel considerably indebted to you," he said, "as do many others in the Country, for your vigorous and successful fight against the restrictions on travel that our State Department was imposing in a rather high-handed way." He went on to tell Nathan about the celloidin-embedding process at the University of Pennsylvania, where, he wrote, "I do part of my work." That he had not worked there in five years was, apparently, an unnecessary detail. Then Harvey offered his first optimistic estimate for the project's completion: "Once we have obtained the [slide] sections, it will take

probably several more months to have them all studied." He said he
hoped to show Nathan the embedded specimen when the executor
next visited Princeton, which, he added, "would also afford us the
pleasure of seeing you." Nathan, apologizing for the delay, thanked
Harvey with a warm letter two months later, saying, "I am sure that
you will be in touch with us when the work is completed."

There was a 1950s formality to their correspondence, both men
referring to themselves as "we" and "us." But the choice of pronoun
also allowed pathologist and executor to leave the impression that
they represented wider institutional interests and not just their
own. Yet neither Einstein's blood relatives nor Princeton Hospital
had had any further input concerning the matter since the week of
the Nobel laureate's death.

✶

While waiting to hear that Keller's work was done, Harvey pon-
dered what he would do next. He would send Zimmerman a set of
slides, of course, if he was still interested in being involved. Perhaps
a couple of those fellows from the Washington meeting. But who
else? Harvey felt he should know, and preferably trust, the scientists
who would receive slides from his amazing specimen, and that nar-
rowed his choices considerably. So when Harvey bumped into
Raymond Stone one day at the hospital, he remembered that the
staff obstetrician had once mentioned having served with a neuropathologist during the war.

Yes, Stone said, that's right. His friend Sidney Schulman was now
practising neuropathology at the University of Chicago. When they
were serving together in Puerto Rico, Schulman had discovered a
treasure of preserved brains taken from the corpses of U.S. servicemen going back decades. The pickled organs filled shelf after shelf,
like mason jars in a pantry. Schulman, he said, had volunteered to
write up descriptions of them.

That Schulman was only a junior, a resident at the university in Chicago, seemed irrelevant to Harvey. He asked Stone if his buddy would be interested in taking a look at slides of Einstein's brain.

Stone was thrilled and didn't hesitate. "Yes, definitely," he said. Then he ran off to call his friend.

Harvey never thought to compile a list of possible candidates to study Einstein's brain from the neurology departments of various universities. Nor did he consider at that point looking beyond American borders, to ship samples to Wilder Penfield at the Montreal Neurological Institute, or any of the leading research centres in Europe. He would be able to say with some confidence in later years that "nobody knew what you could learn from the anatomy of a brain back then." He would comfort himself with the notion that he simply had a specimen that the science of the day was unprepared to deal with. As well, he reasoned that if he had a personal connection with the scientists, they would be more likely to conduct the analysis free of charge. As it was, "I couldn't pay researchers to study it," Harvey said, "and they all had their own work to do."

While brain research was a hot area, hunting a biological source for genius hardly ranked as a priority. For one thing, the Nuremberg trials had revealed the horrific details of Nazi medical experiments, including heinous attempts to prove the Jewish brain biologically inferior. To turn around and investigate anatomical signs of superior intelligence might be construed as more than just poor judgment, not to mention irrelevant, given that there were diseases to cure. War veterans had come home brain-damaged. They suffered blackouts, memory loss and speech impairments, and they complained of phantom pain in arms and legs blown off in battle. Researchers were busy looking for neural defects that might account for it. They relied on pneumo-encephalograms that shot air into a patient's spinal column and up into the brain, where it pushed through various ventricles, highlighting unusual masses or scar tissue. Since the days of Paul Broca, brain lesions had proved to be the most instructive method of finding out the function of the organ's different parts. A nasty bump to the back of a head could lead to blindness. Damage to the crown could keep a person from adding and subtracting. Doctors experimented with psychosurgery, removing frontal lobes, and nuances of personality with it, to calm troubled patients, and extracting various brain regions to curb epileptic seizures.

At the same time, the 1950s offered scientists the first hope of effectively treating mental disease. They'd learned from animal experiments that adjusting neurochemical levels could alter behaviour, making cats more aggressive, dogs salivate or monkeys placid. In 1949, Australian psychiatrist John Cade pumped lithium into one of his patients for five days and reported him cured on the sixth. After twenty years of unsuccessful talk therapy, of uncontrollable babbling in a mental asylum, Cade's fifty-one-year-old patient was able to pack his things and go home. Psychopharmaceuticals flooded the market soon after. In 1954, doctors started prescribing chlorpromazine for schizophrenia instead of a padded cell. In 1955, meprobamate arrived to tranquillize the anxious. Imipramine, the first antidepressant, reached the shelves in 1957. Brain drugs suddenly freed patients from their straitjackets and shackles. Scientists were more interested in the workings of a diseased or normal brain than a brilliant one.

✴

When Marta Keller was done, she presented Harvey with ten pine boxes. Each one was roughly the size of an encyclopedia volume, fastened shut by a tin clasp. Thin slots inside each box held two hundred microscopic slides, stained and ready for viewing. They were magnificent, Harvey thought: "The best I've ever seen." He made his first delivery right there at the University of Pennsylvania. It was not presented as a specimen for scientific study, but as a souvenir for a former boss. Wilhelm Ehrich, a general pathologist, had been Harvey's supervisor when he'd first arrived at the school as a pathology intern. Ehrich had always been very kind to him, allowing him to upgrade his skills with various courses and spend time with Lewy in the neuropathology lab. Harvey did not expect Ehrich to analyze the slides at all. "It was just as a gift of appreciation," he said. It gave Harvey no pause that making a gift of tissue wafers from Einstein's brain breached his promise that the organ would be used only for scientific study. Ehrich was, after all, a scientist and Harvey felt sure that he would not exploit the samples. No one, including the family, ever needed to find out. Ehrich, meanwhile, could see what grand adventures his former apprentice had risen to. Ehrich thanked Harvey and put the slides away in his office, where they would remain for the rest of the century, passed on to the researcher who would inhabit the space long after Ehrich was gone.

Harvey's next stop at the University of Pennsylvania was to the office of Hartwig Kuhlenbeck, one of the few neuroanatomists who had offered him a warm reception at the Washington meeting. Kuhlenbeck was pleased to receive the set of slides. But, he told Harvey, it would be difficult to devote any time to studying them at that moment. He was in the midst of collecting material for a book on the debate around the duality of mind and brain, as well as compiling a multi-volume work on the central nervous systems of vertebrates and invertebrates. Perhaps if the assignment had been passed down through the official channels of Haymaker and the armed forces, Kuhlenbeck would have made the time to study the slides. But as it was, Harvey felt he had no choice but to be patient. "I wasn't going to badger him," he said.

The eight remaining boxes Harvey loaded into his car, along with the two glass jars filled with alcohol and the numbered nuggets of the celloidin-encased "gross material." Since he had the slides, Harvey saw little need to keep the jars at the Princeton lab any more. For a while, they sat on the mantel in the Harveys' living room. At some point, someone, maybe Elouise, moved Einstein's diced brain downstairs to the basement. Once in a while, the Harvey boys brought special friends over to Jefferson Street and took them downstairs to see it. But they were careful not to talk about it too much in public.

"We sort of kept it quiet. He was a pretty famous person and we didn't want just anyone to know about it in case maybe somebody wanted to steal it," Arthur Harvey said.

Despite their discretion, word got out. Louise Sayen said she heard from her son that friends at school had ogled it.

The youngest boy, Robert Harvey, was always a little skeptical about whether he was actually showing off the correct specimen. His dad stored so many pickled parts in his roomy study and down in the basement that he could never be sure. There was the mahogany cabinet from which a real live skeleton popped out to scare the bejesus out of you when you opened the doors. And there was the mummified arm, Robert never knew whose, that he stuffed one day into the sleeve of his own jacket to frighten his friends at school. Robert never thought his dad peculiar for keeping these curiosities. He knew what his father did for a living. Harvey had taken his youngest son to the hospital autopsy suite when he was eight or nine, and Robert had watched with awe as his relaxed father cut open and described the naked dead man splayed out on the steel table. Robert never felt at all queasy. "I remember thinking that my father seemed very proud of his work."

✺

Sidney Schulman lived in a state of suspense for several days. Ray Stone had called him in Chicago out of the blue, sounding playful and mysterious. "I'm going to be sending you a little something," Stone had teased. "I'm not going to tell you what it is. But you're going to find it very interesting."

Schulman didn't have a clue what Stone was on about, until Harvey phoned and spoiled the surprise.

Harvey said he wanted to know Schulman's specific neurological interests. He had slides of the cortical frontal lobe and the parietal region and every other area of the brain, except for the cerebellum and brain stem, which he had not sectioned. But there was no sense bringing everything to Chicago.

"So," Harvey asked him, "what particular part of Einstein's brain do you want to study?"

"I was only a junior at the time and I thought it would be kind of fun to see what Einstein's brain looked like," Schulman said. He wanted to study the thalamus, he told Harvey, from the inner middle-brain.

No one then knew the function of the two egg-shaped lobes of grey matter perched atop the brain stem, but Schulman was intrigued by its complicated anatomy of dense neurons, whose branches radiated outward to the cortex like the spokes of a wheel. Some researchers suspected that it might be important in forming memories, because of a couple of bizarre mishaps in which people had suffered temporary amnesia after a pool cue or fencing foil was accidentally rammed up a nostril and pierced the thalamus. Decades would pass before science identified the structure as the major sensory relaying station, receiving sensations of pleasure, pain, fatigue or shock from the rest of the body and transmitting them to other areas of the cortex to be processed. It was the structure most likely activated when Einstein doubled over in his bathroom, or first clutched his middle at the Brooklyn hospital in 1948. But for all neuroscientists knew at the time, the thalamus might have played some role in Einstein's genius. Or, it might not have.

It never crossed Harvey's mind that the whole exercise was bound to be worthless because no one would be able to conclude whether Einstein's thalamus worked differently from anyone else's. Instead, he told Schulman that he would bring the slides to Chicago soon. He couldn't say exactly when; it depended on when he could get away. Einstein's brain, he explained, was not his full-time job.

It soon struck Harvey that the names of his top picks for the study had been right under his nose since spring: Percival Bailey and Gerhardt von Bonin, the authors of *The Isocortex of Man*, his unwitting instructors. He might have thought it presumptuous to recruit such prominent neuroscientists to work on his project, except that his project involved Einstein's brain. Von Bonin had attended the Haymaker meeting in Washington, so obviously, Harvey assumed, he had an interest. As well, since he'd tried to use their techniques to dissect the brain, he thought he would have the perfect entrée. And if he was going to drive all the way to Chicago, Illinois, to drop off slides to Schulman anyway, what harm would it do to try? He would take his camera, he thought, to snap a few frames of his brain bible's authors. He was anxious for them to see those gorgeous slides. His name could appear right alongside theirs if they published a report: Bailey, von Bonin and Thomas Harvey.

Schulman was out of town when Harvey arrived in Chicago with half a box of slides. He left them with a secretary and said he'd be in touch. Then he drove over to the University of Illinois. Percival Bailey was working in his office when Harvey knocked. Bailey was a husky man with a silvery brush cut and thick, black-framed glasses. He wore a tie and tweed jacket and stood up to shake Harvey's hand. Harvey told him it was an honour to meet him, that he was a big fan. "You did a real nice job on that monograph of the brain," he said.

Bailey called out to his colleague and von Bonin popped in, but he said only a brief hello and then disappeared. Harvey wondered if he'd offended the German scientist back in Washington—or, then again, maybe he was just too busy.

Harvey told Bailey a little bit about himself and how he had been fortunate enough to make arrangements with the Einstein family to study the brain. He mentioned the big-name neuropathologists he'd worked under, like Harry Zimmerman and Frederick Lewy, and Bailey agreed that they were all fine, fine scientists. Then Harvey, careful not to take up too much of the older man's time, proudly presented Bailey with one of the wooden crates.

"I look forward to hearing your observations or your opinion as to whether the slides show any anomalies," he said.

Harvey did not suggest a timeline for the project. Another man might have made arrangements to call in a couple of months to see what Bailey had found. Another man might have asked Bailey to suggest other scientists for the task. But Harvey would have felt as though he were imposing on the eminent neuroscientist. "I wasn't forceful like that," Harvey said. "I think [Bailey] certainly appreciated getting the slides, he was very friendly to me. But he never did say what he would do with them, and I didn't ask."

So there he stood, the man who by happenstance possessed the century's great brain, facing his first opportunity to organize a formal study of it, and he was too timid to discuss any parameters at all. Yet somehow Harvey plucked up the courage to pull out his camera and ask Bailey to pose for him so that he could capture a visual record of the day he came to see him. Bailey obliged. He remained in his office, seated at the round table cluttered with textbooks and towers of papers, smiling into the lens wielded by this curious delivery man.

After Harvey left, Bailey shouted to a young neurologist working down the hall. "Louis, come here," he said. "I'd like to show you something. I know you knew Einstein and I think this will be of some interest to you."

Louis D. Boshes had met the genius physicist while working in Princeton. Boshes had been charged with the administration of a philanthropic fund established by the widow of a clothing manufacturer, and Albert Einstein was one of its advisers. Boshes met him twice, and both times kept his thrill in check. "It was just good manners," he said.

Just how Bailey came to possess sections from Einstein's brain Boshes never asked, but he knew that human artifacts had a way of making the rounds. He had been an intern on night duty at Cook County Hospital in July 1934 when certain organs were pinched from another celebrity corpse. While working overnight in the contagious disease section, Boshes had been called down to the morgue. Men smoking cigarettes and toting submachine guns under their arms were waiting there for him. One of them said he was FBI Special Agent Melvin Purvis. "You better examine the goddamn stiff, Doc," he said. Leather shoes poked out from beneath a blanket and Boshes hauled back the cover. It was John Dillinger, dead as a post. FBI bullets had ripped a mouse-sized hole through the base of the gangster's skull. Boshes phoned the pathologist to conduct the autopsy. But later, before Dillinger could be outfitted in his last suit, five of Boshes's medical student buddies cracked open his head and swiped his brain, swearing Boshes to secrecy.

"So I knew how these things could happen," he said, "for posterity's sake."

Boshes and Bailey both looked at the slides that Harvey had brought, but they doubted anything could be gleaned from them. "Bailey was certainly not going to drop everything else he was doing just to study Einstein's brain."

☼

Updates on Einstein's brain had disappeared from the newspapers following the week of his death. But behind the scenes, countless reporters tried to keep tabs on the research. Harvey had learned his lesson since the Zimmerman squabble. He refused all requests for interviews, refused to pass on the names of other scientists involved in the project and referred the more insistent inquiries to Otto Nathan. The executor enforced the code of silence, turning all of them down flat, even a proposal from the editors of *Life* magazine, who had hoped to publish a photo essay on the brain research. Nathan's position on the matter became so well known that a writer with *Collier's* magazine, hoping to interview Thomas Harvey, phoned Hans Albert Einstein in California for permission instead. The first anniversary of Einstein's death was approaching, and Hans Albert had never heard another word from the man who took his father's brain. Perhaps he, too, was curious about its fate, because the *Collier's* writer sent a letter to Harvey at Princeton Hospital saying that Einstein's son had given his approval for an interview. But Harvey was too cautious to be caught again. He told *Collier's* that he wouldn't say a word unless permission came in writing from Otto Nathan as well. Then he forwarded a copy of the *Collier's* letter to Nathan, explaining: "This seems to me such a reversal in attitude on the part of the Family, that I am very skeptical of the *Collier's* statement."

Nathan, in turn, penned the refusal to *Collier's* on February 10, 1956: "I regret very much that I am unable to authorize Dr. Thomas S. Harvey of the Princeton Hospital to engage in the preliminary interview which you have requested. We have a definite arrangement with the Princeton Hospital that the results of the brain analysis will, if at all, be published in a scientific journal. . . . I am forwarding a copy of this letter to Dr. Harvey and to Dr. Hans A. Einstein in San Francisco, California."

The associate editor at *Collier's* returned the letter saying he was disappointed that Nathan would refuse "to encourage Dr. Harvey to talk over his research project. . . . It seems hardly in keeping with Dr. Einstein's own views on the dissemination of information." The media had, after all, always viewed the genius physicist as public property. *Collier's*, like other news outlets that had been turned away, did not pursue the story. Reporters of the future might have hammered other angles; compiled a biography of Thomas Harvey— probed his qualifications for the job, grilled hospital staff for details of the brain's storage and perhaps even tracked down Marta Keller or the artist who painted its portrait. But television had just begun its invasion of the family living room, and the media had yet to saturate society as they would later in the twentieth century. Before Vietnam and Watergate exposed the sins of the Establishment, the fourth estate was clearly less dogged about championing "the public's right to know." If Einstein's executor and the study leader at Princeton Hospital told reporters to go away, they did—for a long, long time.

Harvey no doubt scored big points with Nathan for deferring to the executor's authority over Einstein's next of kin, though he, like almost everyone else at that point, knew nothing of the rivalry brewing between them. Nathan wrote to thank Harvey very much for his "kind co-operation" and said he looked forward to seeing him on his next visit to Princeton.

Harvey only wished he had something other than media manoeuvres to report to the executor of the Einstein estate. He had driven to the University of Maryland in Boston and dropped off another full box of slides to Walle Nauta, one of the other friendly faces at the Washington meeting, but he had no idea what Nauta planned to do with them. Harry Zimmerman had received his slides and called to say that the brain tissue looked normal to him, that there was nothing special about it, except that Einstein's brain cells appeared to be in pretty good shape for a man of his age. But Harvey wondered if he could trust his old teacher's opinion, given their debate over who would conduct the study in the first place. So Harvey was not about to take Zimmerman's word as the final one.

"I knew what Zimmerman used as criteria, you know; he hadn't studied any geniuses like Einstein," he said. "But he had studied normal brains and he could have written a monograph like Bailey and von Bonin did, on the basis of what normal brains look like. But just looking at it the way he did, just looking at it, I'm sure it did look perfectly normal to him. But, well, that didn't convince me."

☼

Sidney Schulman returned to Chicago from his meeting surprised to find the slides waiting in his office. There were more than one hundred of them, all from the thalamus, nicely prepared, nicely stained. He had only to insert them under a microscope to take a look at the organ's intimate architecture. He never told any of his colleagues, "Hey, look what I've got," but the news spread somehow. Fellow residents in the lab popped in to see and touch the slivers. Schulman himself spent a long time peering at them. But without the comparison to any other brains, or a definitive understanding of the thalamus, he felt his observations were meaningless.

"I really just expected to see normal brain tissue. I don't think we would have known if something looked out of the ordinary," Schulman said. "We hadn't gone very far in reading the structure of the brain then."

But since Harvey had gone to the trouble of delivering the slides and had afforded him the opportunity to examine Einstein's brain, the young resident felt he at least owed him an informal report. So he wrote Harvey a note describing the thalamic tissue. He included measurements on the thickness of grey matter and the variety of brain cells he had seen. The thalamus wasn't like the cortex, with its mass of white matter lying underneath grey matter. The thalamus was all grey matter, dense with neurons.

He mailed the letter to Princeton Hospital, thanking Harvey for the privilege, and even shipped back the box of slides. Einstein's brain took its first romp through the U.S. postal system.

☼

Nathan was getting antsy. With Einstein dead nearly two years and no report forthcoming on his genius brain, the executor wrote to

Harvey on February 5, 1957: "It is quite some time since we heard from each other last. I should be interested in knowing about the progress of your work, whether it has been completed yet or whether you can indicate when you might come to some definite conclusion."

Harvey was clearly unable to give him much of an update because ten months later, on December 4, 1957, Nathan wrote again, his tone still exceedingly polite.

Dear Dr. Harvey:

The newspapers remind me periodically of the brain analysis which you started on the death of Albert Einstein. I do not in the least wish to inconvenience you nor to put any undue pressure upon you. I should, however, be grateful for a brief note, at your convenience, as to how your work has progressed and whether you expect to come to some kind of conclusion.

I hope all has been going well with you. I have not been in Princeton since the fall of 1956 and have hence not been able to keep my promise to call upon you.

Cordially yours,
Otto Nathan
Executor

Harvey realized that he could not stall Nathan any longer. But a part of him worried that if Nathan thought the research was amounting to nothing, Einstein's executor might rescind the permission for him to keep the brain. "Someday we will learn things from this brain that we can't now," he used to say, and because his own ambitions depended on it he believed it.

Harvey had always wanted nothing more than to publish a report on Einstein's brain, preferably one that would present some physiological explanation for the dead physicist's remarkable talents. In the seventeen years since he had graduated from Yale, Harvey's name had never appeared in the medical literature he spent so much time reading, and often admiring. He had come close once, though. Back at Philadelphia Hospital in the 1950s, he and his colleagues at the lab had submitted a paper to the *Journal of the American Medical Association* describing the chemistry of proteins in the immune system after exposure to certain pathogens. The journal accepted the paper for publication, but Harvey was a young resident then, scrabbling to make ends meet with a wife and three sons to support, and he had been unable to afford the association's membership fees. "They took my name off of the paper because I wasn't a member of the association," he said. "I didn't think that was right and so I didn't join the association for a long time after that and I cancelled my subscription."

He considered all sorts of possibilities for publishing a paper on Einstein's brain, even if he had little of substance to submit, which is how he honestly described the situation to Otto Nathan in a letter dated December 21, 1957.

Dear Dr. Nathan:

I was glad to hear from you again but sorry to find that the newspapers have been periodically calling you. They treat us the same and no doubt will continue to do so until publication has occurred.

I had been planning to publish the report in the *Journal of Comparative Neurology*. A few weeks ago, while in New Haven to hear the first memorial lecture for Dr. Peters, I had a chance to talk about the publication with Dr. Donald Barron, the editor of the journal. It is the advertised policy of the journal not to publish any neuropathology. Our report, to be comprehensive, should contain this even though there actually is only a small amount to mention. The pathology in toto will not occupy more than about one page of text and several plates of photographs. My reason for selecting this periodical is that it has published descriptions of the brains of a few other well known scientists. Dr. Barron felt that the pathology would probably be crossed out, so we have been debating about the best method of publication and wondering whether a small monograph might not be the best way to do it.

I have been wanting to come up and give you a report of our work, and wonder whether I could do this on one of the Fridays in January. . . .

With best wishes for a happy holiday season,
Thomas S. Harvey

If Harvey was going to tell Nathan that Einstein's brain looked unremarkable, according to the two reports he had so far received, he wanted to do it in person. Schulman's package from Chicago had raised his hopes, only to dash them again with its inconclusive findings. But he himself was unprepared to accept the verdict. And the fact that Schulman had gone to the bother of writing up anything at all buoyed his spirits. Perhaps, Harvey thought, he could convince this young neuropathologist in training to continue the analysis. He phoned and asked Schulman to do a cell count. Maybe, he thought, there was a chance he would find more than the usual number in Einstein's thalamus.

The request surprised Schulman. The enormity of such a task was overwhelming. Counting cells could only be done one neuron at a time, and how many thousands of them were there in the dense thalamus? There were few shortcuts. The best a scientist could do was limit himself to tallying a specific brain area and then multiplying to estimate an average, the way archaeologists cordon off and dig up a few square metres at a time. But what would be the point of doing it? Schulman wondered. The count would probably fall within the normal range.

"I just don't want to take the time to do something like that," Schulman reluctantly told Harvey. "I think it would be quite uninformative." Besides, Schulman explained, he had his own, more urgent research to do. He was trying to figure out the function of the thalamus by studying the behaviour of rhesus monkeys that had lesions on that part of their brains. Although imperfect, animal experiments offered the best hope for understanding humans.

Harvey accepted his refusal graciously, acting not at all offended that the brain of Albert Einstein should be passed over for the thalamus of a hairy primate. He wished Schulman good luck and sent him a small set of slides to thank him for his efforts.

He sent nothing to Percival Bailey. If either he or von Bonin had reached any conclusions looking at the slides, they never told him. Harvey never saw or spoke to them again. "I was quite disappointed I didn't get some feedback," he said, "especially from Percival Bailey, because I told him I'd been using his great book as my guide."

No one, not even his children, ever learned where Bailey's slivers of Einstein's brain tissue ended up after his death in 1973. Years later, Harvey would still wonder what happened when he opened up his dog-eared copy of *The Isocortex of Man* and out fell Bailey, grinning from the black-and-white photo he had taken.

SIX

RESIGNATION

☀

ON A WINTRY afternoon in January 1958, Thomas Harvey drove to New York to tell Otto Nathan what he'd learned about Einstein's brain. It couldn't have taken more than a few minutes. As he'd said in his last letter, there wasn't much to tell. So far, the organ appeared to be normal, but he explained that some of the leading neuropathologists who had received slides had not yet finished their analysis. In the meantime, Harvey said, he was still undecided as to where he would publish the report. Nathan seemed slightly deflated by the news. "I got the feeling Otto would have felt gratified if we had found some differences in the brain," Harvey would later say. Harvey promised, as always, to keep the executor informed and made his way back to Princeton.

Nathan was likely tiring of bad news at the time. Overshadowing his hope that science might confirm Einstein's genius was his dread that the intimate details of the physicist's private life would soon be bared for public consumption. Hans Albert's wife, Frieda Knecht, was writing a book. Nathan had always suspected that he and Helen Dukas would stand alone in safeguarding Einstein's reputation from prying biographers, but he had not expected family members to be among them. Frieda Knecht planned to portray her father-in-law with full brushstrokes, as a human complete with flaws and foibles. She had in her possession a cache of love letters and far less affectionate notes Einstein had written over the years to his first wife, Mileva Maric. Knecht had discovered the correspondence while she was cleaning out her mother-in-law's Zurich apartment after her death in 1948. Nathan and Dukas, who could only imagine what the letters contained, asked Hans Albert in 1957 to turn them over to the literary archives they were assembling. He refused. Knecht meanwhile submitted the manuscript to a Swiss publisher, hoping book sales would cover Eduard's medical bills, which her husband had taken responsibility for. As Einstein scholar Robert Schulmann explained, there was bound to be war: "Frieda and Hans Albert had the love letters," he said. "But Dukas and Nathan as trustees had the right to publish them. Nathan and Dukas said, 'Turn them over to us and we'll publish them,' and the family said, 'To hell with you; we'll publish and then we'll turn them over.'"

If Harvey's report had disappointed Nathan, it didn't distract him for long. He had a legal war to wage against Einstein's family. For more than a year and a half, the Princeton pathologist heard nothing from him.

✲

In the early days, co-workers and friends would ask Harvey how the Einstein studies were progressing. It wasn't in Harvey's nature to bellyache about researchers letting him down, and even if he had been so inclined, he felt sure that Nathan wouldn't want him discussing the details with anyone. Things were fine, he'd tell them, just taking a little longer than expected. But as the years snaked past, people stopped asking. Some of his colleagues didn't believe he was doing anything at all with it, and that Einstein's brain had instead become his trophy. Harvey's friend Louis Fishman, the hospital internist, felt it was inappropriate to ask. "It wasn't anyone else's business. That brain belonged to Dr. Harvey, it was his," he said. "The family agreed it was and that it could be." Fishman didn't believe there was much chance scientists would recognize genius in the organ: "Maybe if they had another hundred or so genius brains to compare it to, but Harvey did nothing wrong by keeping it, or wanting to save it for science. He was a trained neuroanatomist, you know."

Besides, Fishman thought, Tom Harvey had his own personal cross to bear. He never moaned about that either, and Fishman respected him for it. But the two men often played squash together, and Fishman came to learn about his troubles with Elouise Harvey. She was a moody woman, Fishman said. She could be sullen and then explode suddenly in a flash of rage. The Fishmans used to have the Harveys to dinner, out at their place in Highstown, but then one day Elouise stopped coming with Tom.

The way Robert Harvey saw it, his parents were just two completely different kinds of people. His mother preferred to live quietly, with as little socializing as possible. But with his job, his civic obligations and his earnest passion to do interesting scientific work, his father often drew the whole family into zany adventures. One summer, an Armenian man who had gone to medical school with Harvey pulled up in their driveway with a carload of chickens and nowhere to go. He'd lost his lab in the midst of an experiment to see whether chickens that ate low-fat feed would be spared from hardened arteries. Harvey kept them in the garage, his friend repaying the favour with giant tins of baklava. After a summer of squawking in the shed had frazzled Elouise, Harvey took the chickens out back and cut them open. "That new feed diet turned out to be pretty good for those chickens' hearts," he said.

Another time, Harvey let a skinny Harvard graduate and self-described inventor live in their driveway in his Volkswagen. Robert considered him an original hippie, hair straggling down a shirtless back, raving about "the people catcher" contraption he'd assembled on the front bumper of his Beetle, a net strung between two post beams designed to catch passengers hurled through windshields during a collision. Harvey hired him as a gardener, and Elouise fired him.

Robert suspected that his father had a hard time saying no when people asked favours of him. Harvey even taught hammock-weaving courses at the YMCA after another instructor backed out. But this was not the sort of life Elouise envisioned for herself. She had taken a job as a librarian at Princeton University. She kept house, raised the children and attended meetings of the Delaware Poetry Society, while her husband brought strangers home for her to feed and entertain. And Elouise wasn't like her husband, who did little more than grimace when he was upset. She let fly. She'd holler and complain and send cutlery sailing past his ears. Harvey would say nothing; he often disappeared into his study. He might tug Africa down off the shelf and lose himself in faraway deserts. Harvey would rather vanish than fight back.

☼

Otto Nathan, however, did fight, and win, too. In 1958, a Swiss court ruled in his favour in the matter of Einstein's correspondence. Although lawyers for Hans Albert Einstein and Frieda Knecht had argued that they should be free to make public information that involved their own family, the judges disagreed. Any letter written by Albert Einstein was deemed to form part of the Nobel laureate's literary estate, and therefore fell under the control of the estate trustees Otto Nathan and Helen Dukas. Einstein's son was forbidden to publish any letter his father had written, either to his mother or to himself. Frieda Knecht might have appealed the ruling, but she died suddenly of an aneurysm in October of that year, and the contested letters would remain hidden for the next three decades.

Nathan turned his attention back to the matter of Einstein's brain. Harvey might be content to wait, but the executor opted for a more aggressive approach. In July 1959, he himself contacted the famous neurologist Kurt Goldstein, who had been a close friend of his and Einstein's for many years. Nathan, like Harvey, seemed to prefer to have a personal connection with any scientist who might analyze Einstein's brain. Goldstein, who had devoted his career to treating brain-injured patients and those with psychiatric disorders, had, like Einstein and Nathan, escaped Nazi Germany, emigrating to the United States in 1935. Nathan hoped he might be interested in studying the organ, even though the neurologist was eighty-one years old at the time. "I am sure that Dr. Thomas S. Harvey, the pathologist at Princeton Hospital who was in charge of the work, would be able to acquaint you with the results of his work," Nathan wrote. "In case you should want to consult with Dr. Harvey on the matter, I shall be glad to advise him of your interest, although you might want to communicate with him direct[ly], without any intervention by myself."

A month later, on August 13, 1959, Nathan wrote to Harvey:

Dear Dr. Harvey:

It has been quite some time since you kindly called on me here in New York. As I have not heard from you in the meantime, I assume that you have not reached a final decision as yet concerning the publication of the results of your research on Einstein's brain.

I am writing today to advise you that I recently talked with Dr. Kurt Goldstein, a well-known neurologist of New York. . . . I mentioned your name to him and told him about the work you are doing, in which he showed great interest. In the event he should communicate with you, I wish you would be kind enough to give him access to your work to the extent you consider it desirable.

Sincerely yours,
Otto Nathan
Trustee

In later years, Harvey would say that he had never heard the name Kurt Goldstein, nor could he recall Nathan ever having written a letter about him.

☼

Doctors around the hospital suspected that Harvey was in the midst of a divorce, and if he wasn't, well, he jolly soon would be. Rumours were spreading about their quiet Quaker pathologist taking up with one of his lab technicians. In a small town of watchful eyes, even the hint of a scandal set tongues wagging. Whatever its actual proportions, the gossip eventually reached Elouise, and she marched over to Jack Kauffman's office. No one knew for sure what Mrs. Harvey said, but because of what happened next the staff suspected that she must have grilled the hospital administrator for running a brothel at the medical centre, and maybe even threatened to raise hell if he let the antics continue.

Kauffman had expended considerable energy to build the hospital's reputation in the community, and the idea of having the pathologist's wife running around town smearing the institution no doubt rankled him. But he could hardly lecture staff on matters of morality. Kauffman himself was embroiled in a little afternoon sizzle with one of the nurses. The staff knew that he would sneak away with her at lunch and whisk her off with him on business trips. He even promoted her up the ranks. He couldn't easily play the proverbial pot sticking it to the kettle and haul Harvey in on allegations of a similar indiscretion. Staff members figured that Kauffman needed to find another way to give Harvey the boot. Einstein's brain became the scapegoat.

The administrator began to grumble openly about the way Harvey had handled the famous specimen, making him out to be some kind of slouch for not producing a single report in five long years. Kauffman spoke to the hospital trustees about it, perhaps complaining that Harvey had struck him as an oddball in the first place, that he had made those troublesome promises to Montefiore in New York, that he cooked his Christmas turkey in the morgue, and so Lord only knew what he was doing with that brain. The trustees generally agreed with whatever Kauffman told them. After all, they had hired him to be their eyes and ears. Robert Lewis, one of the hospital pediatricians, said the trustees thought Harvey ought to be dismissed for "his mishandling of the brain and their embarrassment over it. The affair was really the issue, and they knew that, but they couldn't make much of it because of Kauffman."

The staff never bought the subterfuge. Many of them saw it as yet another hypocritical manoeuvre by the WASPy, country-club clique of trustees and their lieutenant Kauffman, exercising power from on high. Lewis said board members tried to run the hospital the same way they exerted their influence over other town institutions, like the local Springdale golf course and the Nassau Club. Einstein had once mocked the local intelligentsia as "puny demigods on stilts," which more or less summed up the doctors' opinion of the trustees. They resented the board members for dictating hospital procedures when not a single doctor sat among them. If a member of the ladies' auxiliary thought the institution ought to advocate natural childbirth, for example, then that's what the board ordered the doctors to do. Lewis remembered an excellent black nurse who was forbidden to wear a uniform for fear of offending the sensibilities of white patients. But if the doctors complained, they risked their own jobs. "They'd call us up on the carpet for everything," Lewis said. "We had to obey their regulations, and they took strict disciplinary action all the time and threatened to take away our hospital privileges." A doctor without hospital privileges could still run a private practice, but if he couldn't send patients to the hospital, the board might as well have run him out of town.

When it came to Harvey's clash with the brass, the staff stuck by him, incensed that the administration was willing to do without a chief pathologist without so much as a word of input from the rest of the staff. "We were very fond of Harvey. He was an excellent pathologist. Everyone liked him," Lewis said. "We were so pleased to finally have a pathologist on staff so no one liked the prospect of losing him."

In his own conscience, Harvey believed he had nothing to apologize for where the brain was concerned. In the five years he'd kept the brain, Kauffman had never once asked him about it, nor had the trustees. "So far as I know they had never really had any real feeling about that brain," Harvey said. "They never spoke to me about it, never spoke to me about publishing anything about it." If he had truly believed their disfavour had only to do with the brain, exonerating himself would have been a simple task. He could have explained his research efforts to date, shared his hopes for future studies of the brain or shown them Nathan's friendly letters proving that he still had the trust of Einstein's executor. But the board members had more than the brain on their minds, and Harvey knew it. And in his mind, it was easier to leave a job he loved than face what would surely become an uncomfortable, messy confrontation. "It was partly my fault," he said. "They wanted me to come in and talk to them and I never did that. I resigned instead, because of the friction between myself and these people, even though the friction wasn't really there. . . . The community got the feeling of ill will between me and Kauffman. I think Kauffman had respect for me and thought I was a good pathologist. I just thought it was better if I left."

While his letter of resignation did not turn up at the hospital for another fifteen years, Harvey did write a request for a three-month leave of absence "due to ill health" in September 1960. According to hospital records, his employee status changed at that point from full-time pathologist to a member of the "courtesy" staff. But like so many issues involving Thomas Harvey, there would linger differing accounts of his departure. Harvey Rothberg, a staff doctor who wrote a book on the history of Princeton Hospital, said the board of trustees gave Harvey an opportunity to resign, but he declined to do so, leaving the board to officially terminate his contract on November 21. "Dr. Harvey was a fine pathologist, but there was criticism of his ability as an administrator and of his personal conduct as well," Rothberg wrote. He said that Harvey had asked them to reconsider, and the medical staff had declared their confidence in their chief pathologist, but it made no difference. The board had spoken. Yet for all the outward concern about Harvey's handling of Einstein's brain, no one thought to ask if the pathologist would be taking it with him when he left.

Harvey never considered leaving it behind. His mission was incomplete, and he had the patience of a good Quaker to see it through. Science might be unable to find anything remarkable in it then, but his faith that the future would tell a different story was as strong as any other that sustained him. He had considered Einstein's brain his responsibility from the first moment he'd cradled it in his hands, and he was entirely unwilling, perhaps incapable, of letting it go. His own elusive dreams were bottled in those two glass jars along with the mystery of genius.

Thomas Harvey left Elouise in the same year he left the hospital. But until he could find a new place to live, the brain stayed in the basement on Jefferson Street. Whatever significance Harvey attached to it, Elouise cursed him for leaving it there. She told her colleagues at the library: "I wish he'd get that damn thing out of here."

SEVEN

THE DOMINO EFFECT

☼

ACROSS THE SPARKLING surface of San Francisco Bay, to a green, hilly outcrop on the east shore, a Venetian bell tower peaks above the tree-tops and stretches down thirty storeys into the Berkeley campus of the University of California. Sunlight drenches its five hundred hectares. Students flick Frisbees in open fields and loll under trees. They wear jeans and sneakers and backpacks slung over their shoulders, sauntering to class along paths too narrow for anything but bicycles and feet to navigate, past a Greek-style amphitheatre, the Moffit Library, housing the first nugget from the California Gold Rush and a high-perched observatory where the romantic park on fogless nights. From the west entrance off University Drive, the Valley Life Sciences building towers in the apron of a grassy court-yard. Up on its fifth floor, in the summer of 1978, graduate students in the neuroanatomy lab of Professor Marian Diamond saw something in an August issue of *Science* magazine that made them reach for the scissors: "Brain that Rocked Physics Rests in Cider Box."

Science is known to break big news in its pages, as when Scottish researchers cloned a sheep. Often the prestigious, peer-reviewed journal reports incremental developments, meaningful to specialists interested in the acrobatics of a particular protein or gamma ray. But this item had universal appeal. According to the report, inside the box, under a beer cooler, was the brain of Albert Einstein. He had been dead twenty-three years by then, gone before the Bay of Pigs and the miniskirt, before the moon landing and the Beatles. But the genius was as fashionable as ever. He glared out from greeting cards, T-shirts and postage stamps from Nicaragua to Mali. He had become a bona fide brand, his image re-purposed as the quintessential smart consumer selling cameras, hair products, beer and women's stockings—things he never would have agreed to do in life. That very year, Einstein fans from New York to India were planning to throw more than seventy parties to celebrate the hundredth anniversary of his birth. And yet, in a nondescript office in Kansas, in an unceremonious carton, his magnificent brain languished. The students clipped the page and tacked it to the lab wall.

Isn't that strange, Professor Marian Diamond thought when she saw it, that Einstein's brain should turn up after all these years.

Marian Diamond once had no trouble laying her hands on fresh brains. She would just jump in her car and buzz thirty kilometres up the coast to the Veterans Administration Hospital in Martinez, California. She had an arrangement with the morgue there back in the 1960s—not its pathologist, but his assistant, a "diener." In the busy autopsy suites of large hospitals and coroners' offices, dieners keep records straight: the identity of the deceased, the apparent cause of death, weight, height and so on. Dieners also perform the messy muscle work of cutting the first Y incision and cracking the breastbone and the skull. At cleanup, they stuff dissected organs back inside a gaping torso, stitch it up for the funeral home, sterilize instruments and swab blood off the floor. The word *diener* in German means servant, but a morgue diener has far more autonomy than this implies. Often spending more time alone with the dead than a pathologist, these assistants can accomplish a great many tasks on their own initiative. Diamond's deal with the diener in Martinez was a case in point: for a carton of cigarettes, the diener harvested brains on the side for her class at Berkeley. "Those brains were beautiful," she said, "fresh, hard and nicely fixed."

As far as Diamond knew, the diener took only the brains of V.A. hospital patients who had donated their bodies to science. Sometimes she had doubts, of course, but the specimens were terribly important for her neuroanatomy students. "Look at the early scientists, they dug brains out of the ground to get at them, and we felt so lucky to have this supply we were told was willed to us," she said. "You weren't mad scientists, but you really were curious . . . when you are curious, you need a brain and you get one, and we didn't think beyond that." She'd make four or five trips to Martinez a term, bringing in twenty brains a year—one fresh brain for every two students. Not even Harvard, she said, could boast a ratio like that.

Diamond must have struck a glamorous pose cruising happily back to Berkeley with a trunk full of brains: tall, slender, with high cheekbones, her iridescent blond hair wound into a sophisticated French twist, a trademark scarf encircling her neck, flapping behind her in the breeze.

Her father had introduced her to science. He was a doctor at the Los Angeles County Hospital. She used to accompany him on his rounds, which is how, at fifteen, she saw her first brain. Through an open doorway she spied it sitting on a table in the middle of a room. She stared at it from the corridor, contemplating the implausibility that such a rounded plasmic chunk could think. Then and there she decided to study them. She would be the only woman in her neurology class in 1948.

The organ never ceased to amaze her. After more than thirty years, Diamond was sure of the fact that no two brains are alike. Scientists trying to map an atlas of the average human brain concede that there is no such thing as a single representation of it. Brains are as different as faces. Just when she thought she could make sense of the organ, a trip to Martinez would turn up a whopper. Like the day her students recoiled in shock after splitting open a brain completely missing a corpus callosum, the nerve fibres that connect the organ's two hemispheres. They quickly called the hospital to find out about the man's mental condition, but the record indicated nothing abnormal. He was a fifty-two-year-old veteran and blue-collar worker who had died of emphysema. They assumed the condition was rare, but brain scans would later show that an estimated 1 percent of the population is born without the fibrous switchboard in their brains. Some are handicapped by it, lack hand-eye co-ordination and suffer severe retardation. Others seem to function quite normally, confirming the theory that the brain can overcome—or at least compensate for—anatomical deformity or injury. Areas of the cortex related to touch, for example, can expand in people who have been blind from an early age.

The adaptability of a brain, its sheer power to remodel itself, intrigued Diamond above all else. When she launched her neuroanatomy career, she wanted to explore how environment and learning experiences shaped its anatomy—unorthodox ambitions in the early 1960s. Many scientists had abandoned the once popular premise that social conditioning moulds human behaviour. Biology had displaced the Freudian influence with the glut of brain drugs that hit the market in the 1950s. Cheaper and more effective, pills did in days what decades of psychotherapy often failed to do. The 1953 discovery of DNA had meanwhile generated a greater sense that people were genetic products born to their fates. So the scientific pendulum swung; biology predetermined behaviour—not whether you lusted after your mother, or your father smacked you as a child or didn't send you to private schools. Mind was seen as a function of the brain, and only hard science—drugs or surgery—could reshape its physical structure. Nurture seemed less relevant than nature. Right there at Berkeley, Arthur Jensen, a professor of educational psychology, was hard at work researching his 1969 landmark paper arguing that genetics predetermined intelligence and that little could be done to raise a person's IQ. He would go on to make the controversial case that blacks must be intellectually inferior to whites because the average black American scored fifteen points lower on IQ tests and no amount of remedial work could bridge the gap.

Diamond's first major project challenged the established wisdom, which fit nicely with the circumstances: she worked during the university's famed Free Speech Movement, when the campus roiled with dissent. While impassioned students at "Bezerkeley" staged sit-ins and protests, orating from the rooftops of police cruisers for the right to advocate political causes on campus, Diamond toiled in her lab, building cities of rats.

The decision to conduct research on the brains of humble rodents was purely practical. Human brains don't arrive with records describing a patient's mental history. Diamond didn't know whether the person whose brain she was studying had been alert and active or bedridden for years. No one, including she, would have known a Mozart from a Manson. So how could anyone possibly associate the multitude of immeasurable forces a person encounters in a lifetime with the physical traits of a single brain? With a rat, scientists can play God. They can control everything: how it lives, what it eats, what it sees, with whom it mates.

Like all cities, the rodent metropolises that Diamond and her students built had a right side of the tracks and a wrong side. Through the course of several experiments, they studied the haves and have-nots. Lucky mother rats lived with their babies in roomy cages filled with spinning wheels, ladders and mazes. In the slums, one mother lived alone with her offspring in small bare cages.

Diamond accumulated one statistic after another, demonstrating that the rats that lived well boasted bulkier, brighter brains. For one thing, they ran a better maze. Rat brains in the impoverished cages actually shrank. Yet even if they spent the first 766 days of their lives in sparse, standard conditions—roughly equivalent to 75 human years—they grew bigger brains and performed better after spending their last 138 days in an enriched cage. Young rats responded best to the stimulating environment. One area of their smooth, marble-sized brains grew up to 16 percent larger than those of the bored and lonely. Adult rats, as well as the very old, also registered a cerebral cortex up to 10 percent larger if they lived in luxury. But unless Diamond continuously filled the cages of the lucky rats with new entertainment, their brain size decreased as they tired of their toys.

Early reports from the studies, which eventually spanned twenty years, positioned Diamond on the scientific map as a maverick, beavering away on the frontier where hard science flirts with psychology. Critics who wrote unkind letters balked at her suggestion that the physical structure of a human brain could change with experience. How could you compare the noggin of a rodent with the complex elegance of the human cerebral cortex? What the devil did size have to do with mental ability? Had not earlier scientists already discarded brain size as a pointless measurement? Remember the genius Karl Friedrich Gauss and the common day-labourer? Remember the mammoth organ between the ears of an idiot?

Diamond ignored the critics and forged on. Using techniques pioneered by European scientists, she carefully shaved bits of her rat brains to find out what specifically accounted for the greater mass of their organs. It did not necessarily matter how large a brain was, she suspected, but how a brain was enlarged. Under the microscope, her students saw that the neurons from the outer layer of the enriched rat brains had sprouted more and longer dendrites across their surfaces than those of the impoverished rats. What's more, she found these branches boasted more spines along their dendritic length, increasing the number of points at which each cell could receive impulse signals from other neurons. The more connections

neurons can make, the greater the brain power, Diamond concluded. But there was something else. Rat brains from the enriched
cages seemed to have more cells in general in the outer shells of
their cortex. They couldn't be neurons. Neurons of the cortex
were not known to multiply after birth and Diamond and her students did not find their number increased. This was another type
of cell. They were glial cells, shaped like stars and blobs, crowd-
ing the space between neurons, filling gaps between dendritic tips
where connections occurred.

Could this be true? Did the smart rats have more glial cells than
the laggards? Diamond realized the task she would have to undertake
to find out. She would have to count the cells in rat brains from each
group. But there were trillions. And even if she hived off a sample
section for counting, that would still leave hundreds of thousands, if
not millions, and they could only be counted one minute glial cell at
a time. When another professor heard what she intended to do, he
thought she was nuts. "Marian, that's going to take you two years,"
he said.

"I don't care, I want the answer," she told him, "and I don't mind
doing the tedious work to find it out."

☼

It was Rudolph Virchow, the father of the slow, methodical autopsy,
who first identified glial cells in 1858. He named them "glia," bor-
rowing the Greek word for glue because he assumed that they worked
as nothing more than connective tissue in the brain, thus relegating
them to the back pages of neurology for the better part of a century.
Yet even when neuropathology hit its heyday in the 1950s, scientists
still suspected that glial cells played only a subservient role to the prima
donna neurons: feeding them sugar, keeping them clean. Glia out-
number neurons by ten to one in the adult human brain, appearing

in the magnitude of 10 trillion and sustaining the ability to divide and multiply throughout a lifetime. But when researchers attempted to record their electrical signals, as they had done with neurons, glia made not a peep. So scientists once again dismissed them as a "silent majority." They were simply nursemaids, dumb and devoted, waiting hand and foot on neurons. Hardly a sexy research topic— who wanted to probe the life of an understudy?

By the 1970s, only a handful of papers had been written about glia. Scientists were still more than a decade from understanding their sophistication: how the star-shaped astroglia help to build a fetal brain; how oligo blob-shaped glia produce the fatty white myelin coatings along axons to speed signals between neurons; how they act as scavengers, cleaning up cellular debris after a stroke, injury or cell death; how microglia are the soldiers of the immune system in the brain; how they produce and deliver chemicals that influence the way a neuron behaves and how, without them, a human dies. Years later experts would call it remarkable prescience that Diamond attached such importance to them long before the rush in molecular biology.

Diamond recruited a slate of part-time researchers—most of them women, anxious to stay current in science while they had young kids at home. "I'll hire a half-time woman any day or two half-time women to work a full-time job," she said, "because they work so hard at half-time jobs."

The demands were literally eye-boggling. To count cells in rat brains, they photographed one square millimetre of cortical tissue and blew it up to one square metre just to inflate the cells to the size of gravel in asphalt. They covered the magnified copy with a plastic sheet and began. One, two, three, four . . . writing the numbers down on the sheet in magic marker to keep track. Then they laid a second plastic sheet over that, and another researcher double-checked the original totals. "Nothing beats the human eye!" Diamond would tell them.

Her incredulous colleague turned out to be right: the project ate up two and a half years. But the results supported her hypothesis: enriched rat brains had a significantly greater number of glia per neuron than impoverished ones. It was Diamond's belief that stimulated brains had busier neurons, and these needed the help of more glial cells to keep them running smoothly. The brain was like a muscle, she felt, growing big and strong with use. She published the work and drew the ire of her peers once again.

Before time and evidence changed his mind, Bruce Ransom, a glia research pioneer at the University of Washington who entered the field about the time Diamond's studies were published, likened her work to pseudo-science. "At the time, I thought, 'This is so stupid.' I thought it was totally premature to create the impression that a creature could become more intelligent because its glial-to-neuron ratio had increased," Ransom said. "I was almost offended by it."

Diamond was not discouraged. Now, whenever she picked up human brains from Martinez, she would eventually have her graduate students start counting glial cells. Teaching took on a more profound and satisfying dimension now that she had a visual image of literally expanding her students' brains with information. She even used to take a pickled brain to her children's classes, carrying it in a floral hatbox and then laying the organ out before their wide eyes on a silver platter, thinking there might be one among them who would fall under its spell, as she had.

☼

It was at one of the international anatomy meetings in 1970 that Marian Diamond first met Gerhardt von Bonin. The world-famous neuroscientist had left the university in Chicago, bidding good riddance to the snow, and moved in the late 1960s to a swank retirement community in Mill Valley, California, where he could smell the ocean.

One day he turned up in Diamond's lab at Berkeley. He was a slight, swarthy fellow with a fringe of white hair encircling his crown. He asked her if it would be all right if he stayed a while and watched the students work. Diamond said it was an honour to meet the co-author of *The Isocortex of Man*, a pioneer of the cerebral cortex.

"Feel free to drop in anytime you like," she told him, and he did, regularly. Von Bonin would stroll around the class, peer over students' shoulders while they dissected, glance into the occasional microscope and offer his own observations. "People don't know what it's like to retire, to suddenly be away from their major interests. He had no lab, no colleagues to talk to, and here I was, and I was no threat, so he essentially hung out," Diamond said, "and I had the thrill of being able to chat with him and share his wisdom."

Von Bonin never told her about the fleeting chance he once had to study the greatest brain of the century. He said nothing about the peculiar meeting he had attended in Washington, D.C., where a Princeton pathologist had refused to hand over the specimen he had cut from the cranium of Albert Einstein. Yet one of the things he did tell her would have a direct impact on the very organ that had eluded him.

✵

One afternoon, Diamond and von Bonin were in her office down the hall from the lab discussing academic questions about the origins of the mind. In the early 1970s, the evolution of the human brain was a hot topic. Researchers had just reported compelling evidence that Darwin was right: our early ancestors probably did slither out of a warm primordial pond three billion years ago. Humans had inherited animal components in their brains. We were lizard and monkey before we were people, adapting to the environment with small additions to our brains over gigantic leaps of time.

The brain of the human species is an amalgam of three brains, one layered over the other, like sedimentary rock on top of igneous.

A reptilian brain huddles deep in the core, where neurons prompt breathing, swallowing, heart rate and mating. One hundred million years later, the limbic brain grew over the primeval stem to monitor the external world of the mammal, gauging temperature, odours, threats and ultimately settling on an emotional state that directed the reptilian brain to beat the heart faster before sprinting for the cave. The cortex came last, allowing higher creatures to interpret emotional stirrings into conscious thoughts, to control visceral urges, to plan and outsmart lesser beings. Its wrinkled grey surface, the cortex, accounts for 85 percent of the brain that makes humans human, enabling us to sing arias, paint the *Mona Lisa*, invent the telephone and fantasize about the properties of space and light.

Scientists presumed that the front lobes of the brain were the most recent to evolve, protruding to raise and flatten out the forehead, endowing people with personality and moral judgment. But the old professor had his own theory. Von Bonin, who had written a book on the human brain's evolution in 1963, told Diamond that he thought the parietal lobes at the crown of the brain, sitting directly above the ears and stretching backward to where the sweatband of a baseball cap would sit, were the latest additions to the human cortex. Wedged as they are at the top of the brain between areas that process visual and motor information, the parietal lobes interpret sensations received from other parts of the cortex, forming, among other things, a sophisticated system of co-ordination that allows people to perceive themselves and the world around them.

Before the dawn of brain-imaging scans, almost all that scientists understood about the parietal lobes came from meticulous clinical observations of patients who had suffered damage to these areas of their brains. At first, scientists assumed that the left parietal lobe was "by far the most significant part of the brain." People with lesions in the left parietal lobe, the brain's dominant half, especially in right-handers, might lose the ability to learn skilled movements, name items, read, write their own names or calculate. But over time, scientists began to document the bizarre abnormalities that affect people with damage to the right parietal lobe. Not only could these patients lose the framework to process the space around them, paying no attention to the left side of their environments but in many cases they lost the ability to perceive their whole selves, denying that certain parts of their bodies even belonged to them. In one case at the Sunnybrook Hospital in Toronto, a nurse in her sixties who had suffered a stroke was completely convinced that her paralyzed left arm actually belonged to an imaginary woman who had stayed in the hospital room before her. In other cases, stroke victims with damage to their right parietal lobes were profoundly indifferent to the entire left side of their bodies. One patient neglected to apply any makeup to the left side of her face, rubbing rouge on her right cheek only, applying eyeliner solely to her right eye and lipstick across the right half of her mouth. The same woman ignored food positioned on the left side of her plate. Other case studies showed that this so-called condition of "neglect" could affect the mind's eye, so that patients trying to imagine a particular place or situation would also envision only half the world, resulting in difficulty mentally rotating objects or imagining objects viewed from unusual perspectives.

Since they governed such complex functions, scientists believed that the parietal lobes were associated with the higher cognitive powers of human "thinking." Such traits convinced von Bonin that the parietal lobes had in fact evolved more after the frontal lobes had formed and pushed them forward. If von Bonin was right, Diamond speculated that the parietal lobes of the human brain might house more glial cells than the frontal lobes. She knew from her rat experiments that there was only one way to find out. So she took off for the highway, heading north to the morgue in Martinez. She collected eleven fresh brains from the skulls of men between the ages of forty-five and eighty and she carved four blocks of tissue from each organ.

Before the human cortex was properly mapped, brain scientists could become as lost as sailors without longitude: with so many bulges and grooves across its surface, two scientists comparing notes could never be completely certain they were discussing the same island of tissue. But confusion of brain nomenclature disappeared with the contribution of Korbinian Brodmann, a bald German scientist with intense eyes and a short, pointed beard. Brodmann had been drawn to neurology by none other than Oskar Vogt, but the two had eventually parted ways and become competitors. Vogt was convinced that the cortex had more than two hundred distinct areas, while Brodmann believed there to be roughly a quarter of that. It was Alois Alzheimer, a few years before he discovered the cellular traits of dementia in 1906, who lured Brodmann to specialize in neuroanatomy. In Berlin, between 1901 and 1910, Brodmann worked like a cartographer, assigning numbers to various brain regions based on their known functions and the structure of cells found in a particular area, and he produced the standard map for generations of future neuroscientists.

Diamond chose Brodmann area #9 of the left and right frontal lobes, and Brodmann area #39 of the left and right parietal lobes for her research. She started counting in 1979, around the same time she married the famous neurologist Arnold Scheibel. They had met six years earlier when her graduate students had picked Scheibel to be their guest lecturer. He and his late wife had published extensively on the links between the mysteries of the mind and the mechanics of the brain, and Diamond had never heard anything like the talk he gave her class. Scheibel had started out as a disillusioned psychology major who went on to train under Percival Bailey and Gerhardt von Bonin in Chicago, more satisfied to look for diagnoses on a slide of cells than the proverbial couch. But it was Oskar Vogt who had been his inspiration. When Scheibel and his first wife were studying in Europe in 1954, they visited Oskar and Cecile Vogt at their hilltop institute in Germany's Black Forest. Vogt was dynamic and officious at seventy-six. With his majestic, bald head and white goatee, he actually looked like a stockier version of the Soviet leader whose brain he had dissected. He wore a white, double-breasted lab coat and regaled the Scheibels with his findings on Lenin's pickled brain and his theories on the cortex, all the while slipping into German and then shifting into French, prompting Cecile Vogt to interrupt: "Oskarrr," she'd say, "in English, please." Vogt showed the Scheibels his enormous library of human brains and pointed out the extra-wide auditory cortex of a Belgian violinist with perfect pitch and the unusual cortical pattern of an artist with a photographic memory. For Scheibel, Vogt's science seemed real, tangible. "The thing that was so dramatic for me was that they showed me that you can find physical evidence in the substrate of the cortex that can correlate to ability." Scheibel went on to spend most of his career at the University of California at Los Angeles, and even after he and Diamond married they decided neither would give up their position. So, alternating weekends, he would fly north to the Bay area or she would fly south.

One summer afternoon in 1981, Diamond found herself sitting in her husband's quiet UCLA office, waiting for Scheibel to return from a meeting, with nothing to do but think. Her mind skipped over the eleven brains she had collected and the prospect that von Bonin might have been wrong. She was discovering that the frontal lobes did house more glial cells to neurons than the parietal lobes, which refuted von Bonin's theory. But the investigation was not a waste of time; she had now collected a database of average glia-to-neuron ratios in two different areas of the human cortex. What could she do with the information? What did the numbers mean without comparison to point to some purpose for the glia being there? Enriched rats bloomed glia like dandelions in their cortices, particularly in the visual cortex of their parietal lobes. But how could she tell if the same was true for people?

At that moment, the random firings of her own neurons seized upon an image—an article hanging on the wall in her lab back at Berkeley . . . a cardboard box, a cider box . . . a chance to study "the most highly evolved brain available in our lifetime." She had seen the clipping a hundred times and thought nothing of it; now she was scouring her memory to remember what it said about Einstein's brain . . . Scientists had found nothing unusual in it . . . But it was out there. But where? Kansas. Something about Kansas. She picked up the phone on Scheibel's desk and called Howard Matzke, chair of the pathology department at the University of Kansas and a former colleague.

"Howard," she said, "do you know anything about this man in Kansas who has Einstein's brain?"

Yes, Matzke told her, he used to live in Kansas but he'd moved. He now lived in Weston, Missouri, and he sometimes toured groups of high school students through his lab.

Almost as soon as Diamond clicked the receiver down, she picked it up again, dialling directory assistance for a Dr. Thomas Harvey in Weston, Missouri.

EIGHT

LOST AND FOUND

☼

THOMAS HARVEY SPENT his boyhood summers in the Midwest. Every June, when school let out, he and his sister travelled with their mother to visit her relatives, the Stoltz family. They caught the Philadelphia train to Chicago and from Chicago continued west to Iowa. Harvey liked to sit near the window and watch America go by, whizzing past factories and new stucco houses, billboards blooming at the roadside and all manner of grazing beasts, happy to see cities disappear into corn fields and silos. He felt safe in the Midwest. Grandmother Stoltz still lived behind the Presbyterian church in Ottumwa where his grandfather had preached. It was a red brick building with white columns and a double-decker porch. His grandmother spoiled him with hoecakes everyday. She would mix milk into cornmeal and sizzle the dough balls on the griddle. Then, with his belly full, he would venture into the "gold oceans of wheat," scouting for roiling storm clouds and the rainbows that were sure to follow. At night, he would unroll his bedding on the second-floor balcony and fall asleep stargazing.

Now here he was at sixty-six. The years had slipped by like fields and fenceposts in the train windows of his youth, and he was back in his beloved heartland. What was it Einstein had said about the illusion of time? It can pass quickly—like an hour sitting with a pretty girl. Harvey's hair had thinned and silvered. Laugh lines sank deep into folds. His eyes failed him at a distance and he wore glasses most of the time. Living in a new state, with a new family and a new occupation, the reinvention of Thomas Harvey was nearly complete, except for the specimen floating in three jars and the wafers wedged in five wooden boxes, the only constants through the metamorphosis.

Then, in the spring of 1978, the phone rang in his Wichita office, reminding him that the world had not forgotten.

"Are you the same Dr. Harvey who worked at Princeton Hospital in the mid 1950s?"

Harvey tensed. He could guess the caller's next question.

✵

Someone, somewhere, was bound to wonder what had ever happened to the most revered brain of the modern age. Brains were making big news in the 1970s. Probing the biology of the mind had become a sexy endeavour. Advances in technology allowed scientists to see a living brain in brilliant detail as the first computerized tomography (CT) scans were performed. The computer-enhanced X-rays captured the organ from different angles to present a single, illuminating picture. As full screens of new clues blipped into focus, it seemed everyone wanted to play detective. Medical students started switching their majors. Membership in the Society of Neuroscience, founded in 1969, swelled to ten thousand over the following decade.

Some of the most fascinating research involved experiments suggesting that the human brain had a split personality. Each half seemed to be outfitted with very different, but complementary, abilities; they were wedded to each other like a biological odd couple. The findings flowed from surgical experiments of the 1950s, when doctors had severed the nerve fibres connecting the brain's right hemisphere to its left in a last-ditch effort to curb epileptic seizures. In 1962, an epileptic war veteran who had undergone the surgery turned up for testing at the California Institute of Technology. The scientists, among them split-brain expert Michael Gazzaniga and Nobel laureate Roger Sperry, found that the veteran had no difficulty naming objects or describing colours or letters flashed in his right visual field—information processed by the left side of the brain. But when they flashed similar objects in his left visual field, waiting for the right side of the brain to respond, the veteran behaved as though he'd gone blind. He swore he saw nothing. Nevertheless, the subject's left hand, also under the control of his right hemisphere, pressed down on the buzzer each time he was presented with a visual stimulus—just as he had been instructed to do. The observation was remarkable. The right side of the brain seemed to lack the left side's ability to label the world with words. Yet the right hemisphere had still prompted the veteran to unknowingly register a visual signal with his left hand. Through the 1970s, researchers showed repeatedly that the right brain was the creative maverick, performing visual and spatial tasks like sketching three-dimensional shapes or mentally rotating objects. The left brain was the logical thinker, excelling at language, speech and problem-solving.

In the lab, scientists sifted through the brain's neurochemical stew, trying to understand the addictions and the biological legacy of the drug-hazed 1960s. In the process, they uncovered the most significant brain discovery of the decade. In 1973, independent research teams in the United States and Europe reported finding tiny receptors lying on the surface of neurons that seemed tailor-made to latch onto opiates. The receptors explained why drugs like heroin and morphine produce such mind-bending effects. But the discovery also raised a far more intriguing question. What were these receptors for? Nature had obviously not supplied them to intensify highs for heroin shooters. There had to be natural substances in the brain that mimicked opiate drugs, and in 1975, scientists announced that they had found them. The bean-sized pituitary gland, on instructions from the hypothalamus, churns out proteins chemically similar to morphine: the brain, in essence, makes its own drugs.

Over the next five years, scientists discovered a whole family of what they called endorphins, proteins capable of improving mood and attitude, reducing stress and pain perception. Researchers suddenly had new methods to treat drug addiction, new chemicals to combat pain and new molecules to mull over when considering the motivations of the human mind. Swedish scientist Lars Terenius, one of the researchers who discovered the brain's morphine-like substances, theorized that endorphins might be the unconscious inspiration for human achievement. He suggested that humans might be the only species on the planet lacking enough chemicals in their brains to keep them happy. Just as people are tempted to drink and take drugs in search of euphoria, so too might they scale mountains, build skyscrapers or pen theories on the laws of the universe if the sense of accomplishment unleashed euphoria-producing brain chemicals. Lower species, meanwhile, would remain content to huddle in their twigs and bushes generation after generation.

☼

While assembling a sixteen-page feature on the mechanisms of the brain in the mid 1970s, *Harper's* magazine editor Michael Aron began pondering the fate of Einstein's amazing organ. Over the course of his work, the subject had cropped up in conversations with researchers. It usually came as an offhanded remark in reference to intelligence and brain size or the speed of brain function. But none of the scientists actually knew whether Einstein's brain had ever, in fact, been researched. The last line of Ronald Clark's 1972 Einstein biography tantalized the journalist. Einstein, the book said, "had insisted his brain be used for research." Had it? Aron wondered. And if so, what had researchers discovered? The book said nothing else on the subject.

Like Galileo's fingers or the shrivelled nub said to be Napoleon's penis, Einstein's brain and its whereabouts had taken on the dimensions of an urban legend. Some of the late physicist's letters and manuscripts had been auctioned off for thousands of dollars, yet there had been no official word about the savant's pickled grey matter in the more than two decades since his death. Rumours circulated that some old codger kept the cerebral masterpiece in his garage; that it was normal and had been tossed out with the trash; that it was frozen and waiting to be cloned.

Aron eventually wrote to Clark himself for information. But Einstein's biographer could tell him only that he knew the brain had been preserved—somewhere. Clark suggested that Aron contact the executor of Einstein's estate for more information.

On March 15, 1977, Aron wrote a polite letter explaining his work and what he had heard about the brain to Otto Nathan's Fifth Avenue office in New York. "I wonder if you would be so good as to tell me what you can about the disposition of Einstein's brain at the time of his death," Aron asked, "and whether any tests or studies were ever performed."

There was a time when Nathan would have done nothing more than rebuff Aron's letter with a terse reply. But by 1977, the eighty-four-year-old executor had twenty-two years of reasons to harbour a certain amount of sympathy for the journalist's curiosity. How many times had he asked the same questions? Where Harvey had once passed on media inquiries to the executor to handle, Nathan now referred the inquiry to Harvey. Perhaps a part of him hoped that a reporter would apply enough pressure to compel Harvey to publish something, at last. Nathan was careful to conceal from Harvey his intermediary role; he never gave Aron an address or telephone number. But he wrote back immediately, saying nothing to discourage the journalist's interest. Aron, he suggested, should telephone Princeton Hospital. If the hospital should be unable to share the results of the research, someone there might be able to tell him where to reach Dr. Thomas Harvey. Harvey, Nathan explained, was the pathologist in charge of the study. Nathan made it clear: anyone wanting to learn about Einstein's brain had to first find its itinerant keeper.

✻

It was Einstein who said that a river's path would become forever more loopy, its outer currents picking up speed at the slightest bend, causing further erosion and continuous curves. Life was a river for Harvey, rushing in unpredictable directions, carrying him haphazardly from place to place on relentless waves.

After leaving Princeton Hospital and Elouise in 1960, he had stayed in town a short while to attend civic meetings and see his sons. But his two eldest had gone off to college, and Robert left for Vietnam at seventeen. Harvey meanwhile worked half a dozen jobs in ten years. For a short run, he and the Fishmans and Cleora Wheatley, a former nurse at Princeton Hospital, tried to run a nursing home just outside Princeton. But none of them was equipped to be a full-time administrator, and Harvey eventually sold his interest. After that he opened a private medical lab in the very same Princeton building where scientists had once worked on theories to split the atom during the war. Harvey thought it an interesting footnote that the brain that had birthed atomic theory was temporarily stored there. But he was always careful not to leave the impression that he attached any cosmic significance to the coincidence: "That's not why I worked there," he'd say. "It was just a good lab."

He next took a job with the New Jersey Psychiatric Institute. While studying the brains of rodents involved in drug-testing experiments, he enjoyed the company of an attractive research assistant. Lisa Scott-Brannigan was an Australian divorcee raising two daughters on her own. Harvey liked her sharp mind, and the idea of an end to lonely nights. They married in 1963 and had two daughters together. They named the elder child Frances, after Harvey's mother, and the younger one Elizabeth.

Not long after Elizabeth's birth, Harvey accepted a position as the staff pathologist at the Marlboro Psychiatric Hospital. It was New Jersey's largest public asylum at the time, a sprawling campus housing more than one thousand residents, from mentally handicapped children to the criminally insane. Many of them lived in rooms the size of closets. But Harvey rarely tended them there. He tested their blood and urine samples in the lab and supervised their final earthly moments at the morgue. Since they were wards of the state, their cadavers were free for the scientific taking. "I tried to study brains whenever I got them," Harvey said, "to give me a sense of what was normal. But in those days . . . we didn't really know what to look for with comparisons." So by day he handled the damaged specimens of the mental patients. By night he admired the diced brain of the century's intellectual hero.

Harvey kept a microscope in the family's prefab cottage on the institution's grounds. Once in a while, he would choose slides from the wooden crates to analyze. Einstein's neural forest never failed to impress him. A few times, he spotted tangled nerve fibres. He didn't know whether they were axons or dendrites, but he figured there were too few of them to suggest the early stages of Alzheimer's disease. Someday, he told himself, he would get around to drawing an atlas of Einstein's brain, a monograph like the one Bailey and von Bonin had produced. But he no longer saw the need to hurry. "I had distributed most of the slides that I had to leading neuropathologists in the country, hoping they would come up with something, and they hadn't, and I thought if these people couldn't, then I certainly wouldn't be able to. So I was just keeping it."

Religiously, twice a year, he replenished the jars with three parts alcohol and two parts water. He coated the rims with Vaseline to act as an air barrier. For an extra precaution, he secured the lids with masking tape, careful to keep Einstein's brain from drying out.

✣

The estate trustees had their own preservation project under way in the decade following Einstein's death. Helen Dukas, who had stayed on at 1 1 2 Mercer Street, kept her late employer's study just as it had been when he was alive. The physicist's books and record collection remained on the shelves and photographs of Einstein's mother and sister still hung on the walls. Einstein had never made any attempt to organize his papers until his meteoric rise to fame, following the 1919 eclipse. His stepdaughter worked as his secretarial assistant until 1928, when Dukas landed the job. Dukas made a point of collecting and categorizing nearly every shred of paper that passed through his hands. She was known to salvage even Einstein's crumpled scribbles from the dustbin. The second-floor room housed all the physicist's papers that had been smuggled out of Germany just before the war, and Dukas organized those as well. After Einstein's death, it all became part of the literary estate. But only the papers that she and Nathan deemed appropriate—those that showed Einstein in the most favourable light—were taken to the Institute for Advanced Study. There, Dukas helped assemble an archive of Einstein's work. Harvard science historian Gerald Holton remembered discovering the diminutive Dukas there shortly after the physicist's death, flanked by twenty filing cabinets deep in the bowels of the building, toiling in the dim light of a walk-in safe, "very much like Juliet in the crypt," he told Einstein biographers Roger Highfield and Paul Carter.

Nathan had meanwhile published a selection of Einstein's essays on pacifism in 1960, under the title *Einstein on Peace*. He spent the rest of the decade expanding the literary estate, tracking down in distant corners of the world letters and documents his great friend had written. And since the 1958 lawsuit against Hans Albert had confirmed his powers of censorship, Nathan built himself a nasty reputation for exercising it. According to Highfield and Carter, Nathan excised sections of letters before he would grant an author permission to publish them. He prevented translations from German into English to keep certain facts away from North Americans. At one point, he and Dukas threatened legal action against biographer Ronald Clark and his sources if they discussed Einstein's divorce from his first wife in any detail. Nathan thought nothing of repeatedly blasting scholars over the telephone for sharing too much information with a writer, or demanding the surrender of Einstein papers another scientist happened to possess. "Nathan was this very acerbic fellow, a real terrier, incredibly protective of Einstein the man and the scientist," said Einstein scholar Robert Schulmann. "He didn't hesitate to fight with anyone who disagreed with him." It was for these actions that Nathan and Dukas earned their reputation as the St. Einstein Brigade.

Yet even after eleven years during which Thomas Harvey had vanished without a word—without so much as a note in response to Nathan's last letter or a forwarding address—Nathan still showed the pathologist the utmost courtesy. On June 24, 1969, Nathan tracked Harvey to a private lab in Freehold, New Jersey, where he had accepted a job after leaving Marlboro.

> Dear Dr. Harvey:
>
> We have not heard from each other for a very long time. An inquiry has recently reached me which recalled our discussions after the death of Einstein and several years later when you were kind enough to call on me here in New York. I wonder what was the ultimate result of your work on Einstein's brain. Have you ever published anything about it and, if so, where? Whatever information you might be able to give me would be gratefully appreciated.

I understand that you left Princeton Hospital. I hope you are enjoying the work which you are doing now.

With kind regards,
Otto Nathan

Freehold is less than two hours' drive from New York City. It must have somehow just seemed easier to get behind the wheel than put down in words the last decade of his life. Harvey took his new wife with him, and as soon as they arrived he was "real glad he went in person." Nathan, he said, seemed happy to have visitors. The aging economist gave the occasional lecture on the perils of rearmament and sometimes attended pacifist meetings in Greenwich Village, speaking out against the war in Vietnam. But otherwise Einstein's executor had turned into something of a recluse. "He was this sort of house-bound fellow," Harvey said. "My wife thought it would be a good idea to take him out, get him out of the house for a drive." The new Mrs. Harvey chauffeured Nathan on a sightseeing trip of New York while Harvey ran errands. Then they met back at the apartment to talk business. Harvey would hear in years to come that other people felt Nathan was tough to get along with—and he had certainly been on the receiving end of the old professor's temper in the past. But Nathan, he said, "was always very friendly to me."

Harvey explained that his professional and personal life had been upended. If he described the Princeton Hospital trustees as having persecuted him in any way, it might only have earned Nathan's sympathy. The executor, too, had once experienced difficulties with the faculty of Princeton University and had left for another teaching position in New York. Einstein scholar Robert Schulmann, considering the strange relationship between Einstein's executor and the man who took Einstein's brain, felt that there must have been some kind of bond between them for Nathan to have let Harvey's omissions slide. "Perhaps Nathan saw Harvey as an underdog like himself, an outsider," Schulmann said. "Maybe Nathan genuinely liked him, liked that he lived modestly like himself and seemed to embrace similar values." Perhaps Nathan found something comforting in the idea that a like-minded Quaker pacifist should possess the brain. Einstein had once said that if he had not been born a Jew, he would have been a Quaker.

Before he left that day, Harvey told Nathan that although his life had become very busy, he would review his files and possibly prepare a report on Einstein's brain. But even as he said it, Harvey must have realized that a scientist rarely publishes about something he does *not* find. The medical journals have never been terribly interested in featuring accounts of investigations or experiments that have failed. And what if he were to say that it was just a normal brain, and scientists later found it to be extraordinary—as he so faithfully believed they one day would?

Two years of silence passed, and then Nathan wrote again in February 1971.

Dear Dr. Harvey:

I recall with great pleasure the visit which you and Mrs. Harvey paid to me about a year and a half ago. I also recall our discussion at the time. You indicated that, although being very busy, you should like to go over your old files . . . an inquiry about the brain from overseas, has brought the matter back to my attention. I should appreciate a note from you at your convenience.

I hope all has been well with you and the family. Please give my warmest regards to Mrs. Harvey. Best regards to yourself,

Otto Nathan
Trustee

Once again Harvey and his wife, this time with their two daughters, drove to Nathan's New York apartment. They chatted amiably and Nathan told Harvey that he and Helen Dukas were signing a contract that year with Princeton University Press to publish a collection of Einstein's works. It was to be a grand project, a literary monument to the physicist's achievements, which he expected would run to a massive forty volumes. At some point, the two men discussed the possibility that the Princeton press might also be able to publish the analysis of Einstein's brain. It would certainly save Harvey the trouble of finding a journal interested in less-than-spectacular results. But it would also fall on Harvey's shoulders to actually compile a learned treatise in the tradition of *The Isocortex of Man.* "I hadn't done enough work in that line of study to do a monograph and I didn't know anyone at that time I wanted to give to," he said.

But Harvey promised to get back to Nathan with printing-cost estimates. Three months after the visit, Nathan had heard nothing from him. On May 21, 1971, Nathan wrote: "You may recall that you intended to enquire about the approximate amount of which the setting and printing of the material would require. If you could do so I should be very glad to have the figures. It might make it much easier for me to obtain the information which you are interested in." There is no record that Harvey ever replied to the letter, and years later, he could recall nothing about the discussion.

When he received a letter six months later from a psychology professor, John Nash, at the University of Hong Kong inquiring about Einstein's brain for a book that would touch on cellular differences betweeen superior and normal brains, Nathan forwarded it to Harvey immediately. "Unless you have prepared something for publication since we met last February," he wrote, "you may or may not want to acquaint Dr. Nash with your results. In any event, I felt it would be the easiest way if you were to reply to him yourself." Nathan must still have been sensitive to the bout of publicity Harvey had faced early in his tenure as the brain's keeper, because he added: "I purposely did not indicate your name in case you prefer to remain anonymous for the time being. In that case I should be very glad to acquaint [Dr. Nash] with any information which you would want me to give him. May I hear from you soon? I hope you and the entire family have been well. My warmest regards to your wife and yourself."

It was a Christmas card Nathan received in response, not from Harvey but from his wife. On January 13, 1972, with his patience at last waning, Nathan wrote to thank her and recruit her to his cause.

Dear Mrs. Harvey:

I wish to thank you for your particularly nice Christmas card. . . . I would like to ask a favor of you. I know how reluctant your husband is to write letters. I am concerned about a physician in Hong Kong who enquired about the result of the examination of Einstein's brain. I enclose a copy of a letter which I addressed to your husband a couple of weeks ago. Could you try to find out what the answer is to my original letter of November 18. A short note from you would be greatly appreciated.

I recall with great pleasure you and your husband's visit and the two children who came with you. I hope you come to New York again and have a moment to spare for a reunion.

With kind regards,
Otto Nathan
Trustee

Harvey finally phoned Nathan on February 18, 1972, saying that he had misplaced the letter from the scientist in Hong Kong. Nathan wrote in his files that he mailed out another copy, but Harvey could not recall ever receiving it. That year, he and the family picked up and moved west. One of his co-workers at the Perth Amboy lab, just a short commute from the small community of Freehold, had told him that a big lab in Wichita, Kansas, was looking for a supervisor. Harvey had come to enjoy the biomedical lab work, especially since it had given him an opportunity to experiment with early forms of genetic testing. He would add hormones to blood samples, spurring white blood cells to multiply, and then interrupt cell division to take a smear slide that would allow him to detect genetic abnormalities and diagnose certain diseases. He imagined that all sorts of opportunities awaited at the Stat Lab—a facility that served all of Kansas, Ohio, Nebraska, Missouri and Iowa. Few of his colleagues saw it as a plum post. But then, Harvey said, they didn't know and love the Midwest the way he did: the sweet, loamy scent that wafted on the air, the land of hoecakes and rainbows and big starry nights. "Having grown up in the Midwest, why, I was willing to go back. I liked the rural atmosphere that you find there, the open country and the corn fields," he said. "I didn't hesitate." Young Elizabeth would remember that her father filled entire moving trucks with his books and magazines and medical literature. She never forgot how he loaded and unloaded the box containing those heavy jars. "Daddy didn't take care of a lot of things in his life," she said, "but he sure took care of that brain."

Wichita pops straight up out of the Central Great Plains, smack dab in the middle of the country, with no mountains or seaport nearby and few trees to speak of. Originally an old cowtown trading post, by the 1970s the city was a hub of the aviation industry, fuelled by local oil patches and the flat, windy terrain that made it ideal for testing aircraft. The Harveys bought a house twenty minutes south, in the booming suburb of Derby. But Harvey never spent much time at home. He worked like a man doing penance. Monday to Friday, he ran the Wichita lab and volunteered to conduct autopsies at local hospitals. He also joined a medical service that sent him to needy hospitals throughout the Midwest on Saturdays and Sundays—

though he'd had not a lick of clinical experience since his Yale days. "On weekends I would moonlight covering emergency rooms in the hospitals around Kansas, doing whatever work there was, just for the experience of doing general practice. It was an additional income and I wanted to do clinical work. You know, with my illness, I was pushed into pathology." Thwarted in his desire to become a pediatrician, thwarted by the limits of science and his own inability to complete an analysis of Einstein's brain, Harvey kept a punishing schedule, still determined to do something noble with his professional life. It was as though he had decided that he could distinguish himself by the sheer number of hours he put into his work.

✺

It was another Christmas card from Mrs. Harvey, in 1974, that informed Nathan of the pathologist's whereabouts—and, by association, the new address of Einstein's brain. The executor, then eighty-two years old, had lately been suffering health problems. A while back he had fallen off a subway platform. But the injuries had done little to dissuade him from playing a lead role in the Einstein papers project. He and Dukas had agreed with Princeton University Press that a science historian should join the project as an editor. They had assumed, however, that they would retain control of the materials to be included in it. To their frustration, the editors they spoke to would take the job only if they were guaranteed editorial freedom. Five years after the project had begun, he and Dukas—compelled by the director of the university press—were still searching for the appropriate person for the job.

Jamie Sayen used to visit the aging Nathan at his apartment near Washington Square to interview him for the book he was writing, *Einstein in America*. "There were wall-to-wall bookcases and papers piled everywhere, books piled on the floor," he said. "It was a dark room, with some of the dark, heavy German furniture. The shades were often fully drawn." Sayen described Nathan as a man who had always taken care of people: as a young man, his elderly parents; in adulthood, Einstein. Now in his old age, there was nothing but the memory of a man to look after. Most of his days and nights Nathan spent alone, surrounded by a scholar's ephemera. "It was no easy task to take care of the legacy of the greatest man of the century," Sayen said. "Was he a pain in the ass to deal with? You bet. Was he detested? You bet. But underneath that he was a kind and humorous man, if he trusted you."

When Nathan replied to the Harveys' Christmas card on January 2, 1975, an undercurrent of loneliness skittered beneath the friendly tone.

Dear Lisa and Tom:

I was as grateful for your thoughtfulness in sending me wishes for the holidays as I was surprised that they came from Kansas. Although we have not seen each other for several years I was sorry to realize that you are no longer living close by. I assume that Tom has facilities similar to those in New Jersey which will allow him to continue his scientific work. If you find the time to write again I hope you will tell me a little about living and working conditions where you are and will also give me some news about the children.

Little has changed with myself. I have fully recovered from the accident and sickness that plagued me several years ago; in fact, I have recovered better than was to be expected.

With best wishes for the New Year and warm regards,
Otto Nathan

P.S. I hardly dare to enquire about the paper which you intended to prepare on the research done on Einstein's brain. If you could find the time and the leisure to do it I still feel that it would be a great contribution.

With the fact of his mortality fresh in his thoughts, Nathan no doubt feared that he might never live to see a report on Einstein's brain. Hans Albert had died after a heart attack in 1973, just sixty-nine years old. And for nearly twenty years, the physicist's defining organ had floated through the discombobulation of Thomas Harvey's life, suspended in diluted alcohol and good intentions. Had Nathan's own desire to see it championed as an organic example of intellectual perfection blinded him? The notoriously impatient and exacting executor had, after all, allowed himself to be put off for two decades with the vague promise that something, someday would come of it. "Harvey was stringing him along and Nathan was desperate to believe that he was associated with a genius," Schulmann suspected. "He was a bully and he was obsessive, but those two things were not necessarily conflicting characteristics."

As for Harvey, his were sins of omission. It would be hard even for him to explain in later years how the clump of tissue that had so defined his life had actually occupied so little of his mental energy. He was incapable of expressing either to Nathan or anyone else that he had come to see himself as little more than the conscientious time capsule that would carry Einstein's brain into the future, where, he was certain, scientists would find within it the keys to unlocking the mystery of genius. "I would have written Nathan if there was something to report," Harvey would say. The patience and passivity that had served him so well as a Quaker had failed him horribly as the study leader of a great brain. But what better qualities for one content to be a curator?

In his drive to work and work, many things fell by the wayside—including his second marriage. In 1975, Lisa Harvey left Kansas with their daughters and moved to Florida. On December 15 of that same

year, fifteen years after he had left, Harvey finally wrote an official letter of resignation to Princeton Hospital: "Having transferred my work
from New Jersey to Kansas, I am resigning my position," he wrote. "It
has been a valued appointment, although a neglected one. . . . My
best to all the staff, Sincerely, Dr. Thomas Harvey."

☼

Though they had the letter of resignation on file, Princeton Hospital
officials told reporter Steven Levy that they had no idea where Harvey
had gone. Levy's editor, Michael Aron, who had moved from *Harper's*
to the *New Jersey Monthly* magazine, had issued the daunting assignment: "I want you to find Einstein's brain." Aron had passed on the
few leads he had, basically what Nathan had written to him in 1977,
but Levy was getting nowhere. He searched library indices for any
scientific reports that might have been published and found nothing.
He asked around the local barrooms and heard nothing. He might
have been tempted to give up entirely, until a co-worker told him
that she had a friend, a medical student, who had once seen slides of
Einstein's brain tissue. That led Levy to a Dr. Moore, who had viewed
the slides with a scientist by the name of Sidney Schulman in Chicago.

Schulman was surprised to hear from a reporter after so many
years. He confirmed that he had seen samples from Einstein's brain.
But he too voiced a familiar refrain—he had no clue where the brain,
or Harvey, had gone. A co-operative clerk with the American Medical
Association, which Harvey had at last joined, finally dug up an address
and phone number for a Thomas S. Harvey, born 1912, working at
1316 Denene Street, Wichita, Kansas.

When Levy called and asked his question—"Are you the same
Dr. Harvey who worked at Princeton Hospital in the mid 1950s?"—
there was a long pause before Harvey said yes. To Levy, the delay
suggested that Harvey had been considering a denial.

Just as Harvey had feared, Levy identified himself as a reporter who wanted to know about Einstein's brain. He wanted to come to Kansas, he said, to interview Harvey in person. Harvey tried to explain his promise and the agreement with the family not to talk. But Levy persisted.

"I was unhappy that he was coming out," Harvey said. "I'd been trying to suppress articles in the lay press."

☼

The media had always treated Einstein with the attention more often paid to movie stars than scientists. They wanted to know how he slept, what time he awoke, what he ate for breakfast. Pressmen roared up the plank whenever his ship docked in America. A couple of them once fell into the harbour trying to scramble on board. If he agreed to a brief press conference, reporters would jump up and down and climb furniture to be heard, hollering their queries in English to the German professor as though sound could break the language barrier. "These men are like wolves," the famous physicist would say, "each anxious to get a bite at me." Indeed, reporters seemed to consider Einstein the first king of the sound bite. The day he died, the Associated Press distributed a collection of his witty quotations. Of his tatty clothes, he explained, "It would be a sad situation if the bag was better than the meat wrapped in it." Of nuclear fission, he said: "The discovery of nuclear chain reactions need not bring about the destruction of mankind any more than the discovery of matches." In his diary, Einstein described one exchange with the press: "The reporters asked particularly inane questions to which I replied with cheap jokes that were received with enthusiasm."

On his first trip to the United States in 1921, Einstein asked reporters who had tailed him to his New York hotel suite why he should be of such intense "psycho-pathological" interest to readers who had no idea what he was talking about. One of them suggested that by unravelling mysteries of the universe, Einstein had brought humankind closer to understanding God's own mysteries. Though Einstein had once said that science without religion would be dull and religion without science would be blind, he thought it more likely that relativity was simply the latest fashion for the ladies of New York. But his popularity was less fickle than hemlines. When he returned to celebrity-obsessed America in 1930, and then returned for good in 1933, he was still the rage. Photographers waited to snap his birthday picture year after year. They staked out a Brooklyn hospital to catch a picture as he left. They chased him up a west coast highway and caught a glimpse of him standing at the shoreline. A female reporter looking for an exclusive even followed his footprints through the sands of the Mojave Desert, only to find Einstein belly up, sunbathing in the nude. And then, in a torrential prairie downpour, twenty-three years after Einstein himself was reduced to silt, a tenacious reporter in a taxi splashed over the puddled roads of Wichita in search of the scientist's last earthly remains.

✴

The biomedical lab stood in a residential end of town. Harvey had a corner office, away from teams of technicians busy on computers and machines that handled human fluid testing. Levy would describe him as a gentle, grey-haired man who still had a spark in his blue eyes. Harvey wore a shirt and tie and carried a pen in his pocket that had three different colours of ink. He sat behind a cluttered desk and spoke, Levy would write, "as if he expected some buzzer to ring and a voice from the heavens to boom down and say, 'That's quite enough.'"

Harvey spoke few words at the best of times. His closest friends, even his own children, found it tough to guess what he was thinking. And when it came to Einstein's brain, words evaporated completely.

He always worried about saying too much, about saying the wrong thing and inadvertently revealing something that might upset the Einstein family—though he hadn't actually spoken to any of the great thinker's descendants since 1955. Now how would he account for the two decades of silence to a scrutinizing eye?

He told Levy that he had been eager to take on the responsibility of keeping Einstein's brain, for what could be "one of my major professional contributions." He said a group of scientists had studied slides he had prepared, and he stressed the difficulties involved in the analysis. "In order to do a study like this," he said, "you have to have seen enough of the normal brain to have a pretty good idea what would be extraordinary. Unfortunately, not a lot of brains have been studied completely. Less than a dozen. Of course, when it comes to genius . . . not even that many. It really is a mammoth task. There's a tremendous number of cells in the brain. You don't examine every one of them in detail, but you look at an awful lot of sections . . ."

There had been no urgency to publish, he added, and then he offered a vague glimmer of the personal angst that had delayed the project. "You see, my career since I did the autopsy has been sort of interrupted. I left Princeton Hospital in 1960 . . . and for the past few years, I've been here in Wichita. I don't work on it as much as I used to. But we're getting closer to publication. I'd say we're perhaps a year away." He must have known his words might come back to sting him later, make him out to be a stupendous procrastinator. But perhaps if he said it out loud, he might actually summon the gumption to make it happen.

He told Levy that the studies done so far had suggested that the brain fell within the normal ranges of other brains of its age in terms of weight and cellular structure. He didn't mention that most of the scientists who had received samples had never even reported back to him, or that what he had heard he saw no point in publishing. Nor did Harvey tell the reporter that he hoped the brain might yet prove to be extraordinary, that it might yet validate his lifelong commitment to it and all the ambitions he had pinned upon this post-mortem tissue, even while he drifted from job to job, wife to wife, place to place.

But Levy was less interested in the scientific details then he was in the matter itself. He sensed that he might be close enough to lay eyes on the brain of all brains. It was the real purpose of his visit, as it would be for a parade of pilgrims who would come after him— to see, and perhaps to touch, as though Einstein's brain were like a Blarney Stone that could bestow wondrous powers of mind or magic upon believers. Harvey never blamed people if they sucked in their breath the first time they saw it. He never trotted the brain out for that purpose, but there was some comfort in sharing his appreciation of it. A treasure was sunk in the bottom of those jars —the physical remnants of a mind that had spawned nuclear energy and laser beams and weaponry that, at that very moment, still kept the world suspended in a Cold War.

Levy asked about photographs and Harvey had to tell him there were none in the office. Then, Levy would write, "A shy grin came over his face. 'I *do* have a little bit of the gross here . . .'"

Harvey could not resist showing him. Later he would say, "He had come all that way . . ."

He rose from his desk and rearranged a stack of cardboard boxes in the office corner, hauled a red plastic picnic cooler out of the way, lifted one box off the top of another and pulled at the one underneath it. The one labelled "COSTA CIDER." It had no lid, just a blanket of crumpled newspapers on the top. Harvey plunged his hands beneath them and grasped one of the two glass canisters, and then the Mason jar of unsectioned material, pulling up one breathtaking collection of pieces after another.

✧

"My Search for Einstein's Brain" appeared in the August 1978 edition of the *New Jersey Monthly*, along with pictures of the famous physicist and a headshot of a distinguished Harvey, looking not unlike the television doctor Marcus Welby. Harvey thought that Levy had done a fair job, except that he had compared the celloidin-encased brain pieces to "Goldenberg's Peanut Chews." Not quite the respectful tone Harvey had hoped for.

The media that descended upon Harvey as a result of Levy's account gave him confidence that he could handle press inquiries without technically breaking his vow that he would save the science of the brain for scientific journals. Reporters flocked to Wichita. They phoned him at his home and workplace. They sent photographers to snap his picture, and Johnny Carson cracked wise about Einstein's brain turning up behind a beer cooler in Kansas.

"I felt I could talk to them about my role but not discuss the science . . . to keep the science for the journals as the family had requested." He found some satisfaction in knowing that, all these years later, the world still wanted to know if anything could be gleaned from the pickled organ. One neurobiologist wrote to him all the way from the national university in Mexico City to ask whether Einstein's brain showed microscopic evidence of aging. Harvey could not recall if he ever wrote back.

In the end, developments in science itself had rekindled the media's interest in Einstein's brain. And in turn, the media had put the amazing organ back in the radar of researchers. The journal *Science* devoted half a page to the story under the headline "Brain That Rocked Physics Rests in Cider Box." But not long after the Berkeley students pinned the article up in Marian Diamond's university lab, the peripatetic Harvey moved again, to another state, with another new family, in yet another incarnation.

NINE

FOUR TRIPS TO CALIFORNIA

☼

SOMETHING MOMENTOUS happened whenever Einstein went to California. On his first trip, he docked in San Diego at dawn, New Year's Eve, 1930. He was fifty-two years old, touring on the international lecture circuit, invited as a visiting professor to the California Institute of Technology. He never brought his violin—he worried that the climate might warp the wood—but the golden state of glamour and surf offered entertainment enough. Einstein and his second wife, Elsa, toured three different movie studios. They ate lunch with Charlie Chaplin and were invited to dinner at his Beverly Hills estate. They attended a seance and the premiere of Chaplin's film *City Lights*, and later they vacationed in Palm Springs—"loafing in paradise," Einstein called it. But he still found time to work. He was hunting for the theory to prove that gravity, electromagnetism, light and electricity were different forms of the same thing—a unified field theory. The problem had by then obsessed him for fifteen years. "Many of my colleagues think I am crazy," he said. He hoped California's scientists might contribute to his efforts. But instead of making progress on a new theory, Einstein accepted their evidence to discredit an old one.

Back in 1917, he had put forward the notion of a "cosmological constant" as an addition to his general theory of relativity. In mathematical terms, the constant represented a force of nature that would offset the pull of gravity, preventing the universe, in Einstein's model, from caving in on itself. Such a force was necessary to support his view—the prevailing view of the day—that the physical structure of the universe was a static and finite galaxy. But by 1930, astronomers' reports suggested otherwise, and there in California, on the perch of the Mount Wilson Observatory, Einstein changed his mind. Astronomer Edwin Hubble, the Columbus of the skies who would prove that millions of galaxies exist beyond the Milky Way, told Einstein that he had seen with his own eyes distant galaxies blazing away from Earth, gaining speed the farther away they travelled. Hubble gave Einstein a turn at the 2.5-metre snout of the world's largest telescope to peer into the heavens he so often contemplated. Einstein glimpsed the galaxy-studded cosmos for the first time and knew that Hubble was right: humanity occupies a dot in an ever-expanding universe.

Einstein had long known his invention of a cosmic force was a troubled theory, as unstable as balancing a marble atop a billiard ball. But after his night on Mount Wilson, the physicist realized not only was it unworkable, but unnecessary. No cosmological constant was needed if the universe was not static. Einstein eventually called his theory of the cosmological constant "the biggest blunder" of his life. Meanwhile, the world learned that an Einsteinian theory could be overturned like any other scientific apple cart. Still, physics would debate the role of such a theoretical force long after his death, the same way they would continue to question whether he was wrongheaded or prescient in his obsessive search for a unified field theory. In California, Einstein foresaw the mysterious legacy he would leave the world. "I offer it to you like a closed box," he said of his unfinished idea, "and as one who doesn't know what is in it."

✸

Arnold Scheibel was confident that they would find something of interest in Albert Einstein's brain. When his wife, Marian Diamond, told
him about the article that students had hung in her lab, and the call
she had made to the doctor in Missouri, Scheibel said that he too had
once heard something about the famous organ, some newspaper item
"about a fellow who kept in it a pail under his bed or something like
that." Scheibel thought of Oskar Vogt and the German neurologist's
lament that the world had too few gifted brains to study. Now here
was a chance to peek at the cerebral tissue of the century's great genius.
"If I ever had a hero or icon, Einstein was it," Scheibel said. "We thought
it would be fun to try and get some of his brain." Scheibel hoped to
acquaint himself intimately with the architecture of Einstein's neurons. He wanted to study the fine details of the dendrites sprouting
from the brain cells, count them, describe them, record traits that
might help explain Einstein's brilliance. He had already chronicled
the way age, like an avid gardener in autumn, can prune the dendritic
branches of cells in an elderly brain. And, like Diamond, he had
described the anatomical effects of learning. But he realized they
would never be able to say whether Einstein's cells reflected good
genes or good experience.

If Einstein's whole brain were available, Scheibel thought, the organ
could serve as its own control group. Maybe he could compare neurons in those areas of the brain believed to be involved in Einstein's
mathematical and conceptual genius to other areas controlling those
functions in which he was not known to be exceptionally talented.
No definite plans could be made, though, until he determined the
state of the tissue, and he couldn't do that until it actually arrived.
But who knew when that would be? Every few months his wife wrote
to Thomas Harvey. Sometimes, she would call. Each time, Harvey
told her the same thing: that he would send her pieces of the brain,
that he was shipping it soon.

❄

Harvey did not like living alone. He had moved from his parents' home to the dorm at Yale, to the tents of Gaylord, to a house with Elouise, and then within three years of leaving her he was ensconced with his second wife. Einstein once quipped that "marriage is but slavery made to appear civilized," but for Harvey marriage was security, a sense that the world was as it should be as long as there was someone to come home to. After the divorce from his second wife, he telephoned Rachelle Ross, a woman he had known as a neighbour when he'd worked at the Marlboro Psychiatric Hospital back in New Jersey. She was a flight attendant, divorced from her airline pilot husband and raising three young girls when Harvey called. "Well, why don't you come out here?" he said. As he remembers it, "She hesitated for a while, but then she said yes, and decided to come." They married in 1975 and Harvey became husband and father and bread-winner to his third family at the age of sixty-two.

Harvey and his new wife weren't in Kansas long before he received the second most exciting opportunity of his professional life. There was an old doctor in Weston, Missouri, fifty kilometres northwest of Kansas City, who used to send his specimens for testing to Harvey's lab in Wichita, and he was planning to retire. He called up to ask whether Harvey would consider moving to Weston to take over his practice. Harvey had nothing to consider. His whole career he had waited to hang a shingle outside his own office, to be a physician taking care of live patients. It was why he worked the weekend shifts at the hospital emergency wards. And so, at an age when he might have considered retiring himself, Harvey became a general practitioner. For the first six years after he arrived, he was the only doctor in a town of some 1,300 souls.

People sometimes call Weston "Sin City" because it was built on alcohol and tobacco. The oldest distillery in the United States still

stands there, dating back to the 1850s, when a savvy butcher figured
that wagon-wheelers heading west needed whisky more than steak.
At the time, Weston was a port hub kneeling on the east bank of the
Missouri River. But after two bad fires, five floods and one civil war,
most of the population had disappeared. Even the river had receded.
Einstein would have been hard-pressed to find a more obscure place
to keep worshippers from his remains. But despite the empty store-
fronts and the old wooden tobacco barns in disrepair on the hilltops,
Harvey happily called Weston home. He and Rachelle—or Raye, as
he called her—rented an elegant Victorian house and carved out the
same sort of life he had begun in another historic municipality a quarter century before. He doctored at a clinic on Main Street. He joined
the Masonic Lodge and was elected president of the local chapter of
the Rotary Club, spearheading fundraisers and community service
projects to reinvigorate the town. He volunteered as physician to the
high school football team and spent time with students interested in
science. Sometimes he toured them through the anatomy lab of Howard
Matzke at the University of Kansas. He remained quietly ambitious,
his mind scattered across dozens of time-consuming commitments,
none of them involving the 240 pieces of a certain cerebral cortex.

Townspeople liked and respected Harvey. They remembered the
shy, unassuming doctor years after he left—a thin, wiry figure toting his medical bag through town, shuffling past Owen's Barber Shop
and the Rumpel Hardware & Furniture Company to make house calls
on foot. The locals considered him something of a wise man who kept
to himself, even when he popped into the pub to buy the occasional
case of Erlanger beer. They remembered him because, as with Buffalo
Bill Cody, the Wild West hero who had once been a local, Harvey
brought a touch of fame to tiny Weston. He was the man who kept
Einstein's brain. Everyone knew it.

"He didn't toot his horn about it or anything. I think he tried to
keep it pretty quiet. But people sort of knew anyway," said Virginia

Faris, who worked as Harvey's nurse in Weston for several years. "I saw it. It was just there in the clinic in jars, I didn't worry too much about it." She wasn't particularly surprised that Harvey should have come by the famous organ. She suspected that his curiosity had led him down all sorts of avenues of study and adventure: "You always had the feeling he was interested in doing all different sorts of things, that he wasn't just a run-of-the-mill doctor."

Some of the townspeople could be curious too, but few were bold enough to flat out ask Harvey about it. Like the Princetonians who gave Einstein his privacy, Weston's residents preferred not to pry. His landlord, George Rumpel, tried once to find out more about it, but he remembers that the doctor grumbled, uncharacteristically, "That's none of your business." Usually Harvey changed the subject or let the question dissolve into an impenetrable silence. Nurse Faris thought he "would have just as soon have everybody forget about it."

Well, not everybody. Sometimes he would take a few slides or photographs to show the kids at school if a teacher asked him to give a talk. "They're usually satisfied with just seeing a piece of it," he told the *Columbia Daily Tribune*. "I don't have to go into scientific detail." Harvey also shared it with his fellow Rotarians. The club held regular meetings at Bowman's Country Cupboard restaurant, and each week one member was responsible for making a presentation to the rest of the group. When it was his turn, Harvey strolled in with Einstein's cerebellum, a hunk of cortex and the spaghetti-like remnants of the organ's blood vessels in a Mason jar. "It just looked like a brain in a fruit jar," said Rotarian Joe Collison. "But it was pretty impressive, just one of those strange things, I guess." Harvey told them how Einstein's corpse had lain in his morgue that morning in 1955, and how the family had told him that he could keep the brain and study it. But, according to Collison, Harvey also told the Rotary Club one other odd thing: "He said the family didn't want him to release any information about it until after he was gone . . . until after he had passed on. Dr. Harvey said that was so no one could argue about his findings of what the study had to say."

The terms of his promise to keep the brain away from publicity seemed to fade from Harvey's mind the farther he travelled in space and time from its origins. In the interim, possessing the genius relic had indeed conferred a kind of power upon him—the power to hold court and amaze, to keep a class of schoolchildren or a clutch of civic-minded men transfixed. It was this temptation that Harvey resisted less and less as the years passed. He justified it by explaining that the show-and-tell to the Rotary meeting was more an educational seminar than publicity: "People had questions about how this brain might be different from other brains." Questions that Harvey, of course, could not possibly answer. But in a small pond like Weston, Harvey was a big fish.

✹

Despite his best efforts, Otto Nathan was also losing control with the passing years. He and Helen Dukas had, in 1976, finally hired John Stachel, a Boston University physicist, to edit the massive Einstein papers project under way in Princeton. Stachel arrived in January 1977, but within a few months he, too, wanted independence in determining what materials could be included in the collection. Nathan decided that the only way he and Dukas could wield any influence was to demand that the project have more than one editor. Robert Schulmann, the Princeton professor of German history who worked for Stachel on the project after 1981, said that "Nathan wanted a triumvirate to be in charge of the papers and their publication, of which he would have been a member." But the Princeton University Press sided with Stachel, predicting that co-editors—particularly if Nathan were among them—would only clog a process that was already several years behind. "Nathan found himself in the throes of a tremendous fight with Princeton University Press," Schulmann said. The contract dispute went to arbitration, and in 1978 Nathan lost. "As we say down here," Schulmann explained, "Nathan was majorly pissed." But though he was then eighty-five years old, Nathan refused to give up. He hired lawyers and appealed the arbitration decision to the courts of New Jersey.

The last thing Nathan needed was the painful distraction of Einstein's brain turning up in the press—in a cider box. According to Jamie Sayen, it turned Nathan's stomach to think that the physical remains of the man he so loved and had so tirelessly defended were tucked under a beer cooler. "It was supposed to have been an object of serious scientific study, and then that New Jersey article came out and the whole thing just seemed so tawdry," Sayen said. "He realized it was not under any kind of rigorous study." Nathan would have probably been horrified to know that he had actually played a role in connecting Steven Levy to Thomas Harvey. But having no idea that Levy's editor, Michael Aron, had indeed come to him for information, Nathan instead complained that he was "sorry that neither the author of the article nor the magazine felt the desire to be in touch with me before publication . . ."

As soon as Einstein's brain had made its afterlife debut in the press, Nathan's mailbox once again filled with queries. One letter arrived on December 11, 1979, from a Japanese math professor at Kinki University who had just completed a summer tour of "Dr. Albert Einstein's European haunts." Its tone and contents epitomized the fanatical idolatry Einstein had so feared on his deathbed.

Dear Dr. Nathan,

I have some requests to make of you.

Do you know about the publication (exact date, journal or magazine's title, publisher's name, and so on) of the results concerning the "The Study of Dr. Einstein's Brain by Dr. Thomas Harvey"?

Please tell me about the results concerning "The Study of Dr. Einstein's Brain by Dr. Thomas Harvey" by airmail at a rush, in detail. In addition, please tell me about Dr. Thomas Harvey's exact address by airmail as soon as possible.

I'm looking forward to your immediate reply for this letter by airmail as soon as possible. Please do your best for me. With best wishes.

Very sincerely yours,
Kenji Sugimoto

Nathan wrote back a week later: "I do not know of any publication by Dr. Thomas Harvey about his study of Albert Einstein's brain. As far as I know, he has never produced a publication on this subject, but has talked to newspaper people about it. They may have published articles which are not known to me nor do I know Dr. Harvey's present address." This time the executor was not being coy. Harvey had sent him no forwarding address. He had heard nothing from him since Mrs. Harvey sent the Christmas card from Wichita in 1974. When he replied to a query letter from the *Wichita Eagle and Beacon* newspaper in 1979, Nathan carefully tainted the media's impression of Thomas Harvey. "It is correct," he wrote, "that the family and myself had acquiesced when told by Dr. Harvey that he had already eliminated the brain in the morning without permission of the family."

✼

As he steamed across the Atlantic in 1931, California-bound for the second time, Einstein grew frustrated. He was trying to learn English but found himself constrained by the limitations of any other mortal. "It doesn't want to stay in my old brain," he wrote in his diary. During the voyage he had decided that he would likely leave his academic position in Berlin, and English would be a necessity if he moved permanently to a university in either North America or Britain. Germany was rearming itself, anti-Semitic policies were spreading, and Einstein, as he had in his youth, felt inclined to abandon his birth land.

Robert Milikan, director of the California Institute of Technology, was especially hopeful of luring the Nobel superstar to a permanent position in Pasadena. A physicist himself, Milikan had won a Nobel Prize in 1928 for accurately measuring the electrical charge of an electron. He had also spent a decade in his early years trying to solve the mystery of light—how waves of light could act like particles— before the young clerk from a Swiss patent office beat him to it. If Milikan had any reservations about recruiting Einstein, it was only because the professor insisted on publicly involving himself in sticky political causes well outside the sphere of science. In one of several examples of Einstein's crusades for justice, the physicist wrote to the governor of California to free labour activist Tom Mooney and his so-called accomplice, who were serving life sentences for planting a bomb that had killed several people during a parade. Many people believed that the men had been wrongly convicted on perjured testimony, and Einstein wrote: "I myself do not claim my opinion is correct but I think a miscarriage of justice has taken place . . . and you would do much for real justice if you commuted their sentences." As Einstein biographer Denis Brian recounted, California Governor James Rolph leaked Einstein's letter to the press, and Milikan fretted about the possible reaction from Caltech's conservative patrons. Einstein explained that he had not intended his plea to become public.

While Einstein was keen to move to California, he ultimately declined Milikan's offers because the director had refused to give his assistant mathematician, Walther Mayer, a job at Caltech as well. In the end, none of that mattered, because on that second trip to California Einstein met Abraham Flexner. A distinguished educator who had compiled a landmark report in 1910 that improved standards in medical schools in Canada and the United States, Flexner told Einstein about his plans to create an institute for advanced study. Flexner envisioned a place where great scientific minds could spend boundless hours thinking great thoughts, unfettered by the lecturing demands other universities imposed. The idea intrigued Einstein immediately and set in motion his eventual move to Princeton, where the Institute for Advanced Study would be established and where he would spend the last twenty-two years of his life, before ending up on the steely altar in Harvey's morgue.

✷

Autopsies were about the only thing Harvey did not do in Weston. He made house calls around the clock. He covered weekend shifts at the emergency departments of nearby hospitals. Sometimes he flew back to Wichita if a Kansas ward needed an extra pair of hands. He also treated patients at nursing homes in the area. On top of all that, he accepted a part-time position twenty kilometres south at Leavenworth Prison, one of the toughest jails in America. With his characteristic understatement he explained, "I saw the inmates. They'd rape one another and I would have to examine them to see if there was any evidence of the rape." Harvey never complained about the demands on his time. He accepted them, in some cases sought them out, as if to eat up every possible free moment he might have had to turn his attention to the pickled matter that distinguished him. As his wife Raye would one day tell a writer from *Harper's* magazine, "Really, Tom wasn't doing anything with it . . ."

In the ten years Cheryl Schimmel managed Harvey's clinic in Weston she never once heard him bellyache about the hours he kept, never once heard his voice rise in frustration or anger. None of it was work to him, she said. She thought Harvey was "the kindest, most generous and gentle man I'd ever met." She knew the private things he did for his patients. Many of them were much closer to the end of their lives than the beginning and got confused taking their medication. Harvey would trundle over to their homes at five o'clock in the morning. He would lay their pills out on the table for the day, she said, arranging them in separate piles so they knew which ones to swallow when, so they would not have to leave their own houses for a nursing home. Unlike other doctors, Harvey never hired a collection agency to squeeze money out of patients who couldn't pay, Schimmel said, nor did he turn the poor away. At the end of some weeks, he would forgo his own salary to pay his nurse and office manager. To hear Schimmel tell it, Harvey lived the humble life of a picture-perfect Quaker, a charitable humanitarian not unlike Einstein himself. The two men certainly had that in common: a tendency to throw themselves into their work at the expense of all else, to spend more time rushing to the aid of strangers than their own families, to care more for their professional pursuits than the "merely personal," as Einstein put it.

When the *Kansas City Times* interviewed Harvey in 1981, reporter Chris Szechenyl asked why the man who possessed the brain of the scientist who revolutionized our concept of space was working as a country doctor. "That's the way I make a living," Harvey said with a nervous smile. "You don't make a living studying Einstein's brain." Schimmel thought he probably could have. "People from all over the world would call our clinic on a daily basis asking about that brain. He wouldn't talk to most of them. He gave the Einstein family his word and he was as good as his word. . . . *Good Morning America* called and *The Today Show* and all sorts of others, from Japan and from Australia. They all offered him good money to come on TV and talk about it, but he wouldn't even take their calls. He just wasn't interested in things like fame or money."

Harvey must have been a soft touch for reporters from the small Midwestern papers, who usually did manage to land an interview.

But generally they left disappointed. He might have been willing to share the brain story with schoolchildren and the fellows from the Rotary—people who would never second-guess his motives or scrutinize his actions—but not with the media. He offered the press vague answers, saying little about where the brain was being studied or exactly what aspects were under examination. He never pulled the jars out again as he had in Wichita. Who knew what description would turn up the next time? "I always tried to be careful with the press," he said. "I was always loath to give out names or details."

Harvey's ambition to be a man of science had pushed him to take the brain, and yet, ironically, science was the very thing he had vowed not to discuss publicly. In turn, his silence made his status as a scientist highly suspect, so that the only newsworthy tidbit left to chronicle was the bizarre tale of an anonymous doctor and his amazing possession. When Nicholas Wade called from the prestigious journal *Science* in 1981 to ask what his eventual report would be likely to conclude, Harvey went so far as to say that he had "no concrete plans—I have my ideas about it but they have not solidified." The results from the specialists who studied sections of the brain, he said, showed that everything was "perfectly within normal limits except for the changes due to age." But otherwise, his answers in the lay press seemed like the evasive shrug of a schoolboy who hasn't done his homework. "Well, they thought he didn't know anything because he would go all quiet and not say much," Schimmel said. "But that's only because he doesn't talk much if he doesn't know you too well, if he doesn't entirely trust you. But he knew a heck of a lot and, oh boy, he could spit out big words with the best of them if it was appropriate. He had box after box of plaques and awards for the various good things he had done and accomplished but he wouldn't even let us hang his degrees on the wall in the clinic, you know, the way other doctors do. He just didn't want to seem boastful about it, so they just sat there in boxes." Like nearly everything else he owned.

✷

Nathan had now come to rely upon the media he had once rebuffed. Like iron filings to a magnet, they were drawn to Thomas Harvey, and only through them could the executor keep tabs on the wandering pathologist. It was a reporter with the *Kansas City Times* who called Nathan in 1981 and told him that Harvey was working as a family doctor in a little Missouri town. That's when Nathan's anger got the best of him. In setting the record straight and asserting that Harvey had violated Einstein's dying wishes, Nathan also contradicted the very statements he himself had made to the press twenty-six years earlier. "I am absolutely certain that Einstein would *not* have wanted his brain examined. He was too humble," he said. Yet, in another breath, Nathan revealed himself: "I am disappointed that Dr. Harvey hasn't published anything on the brain." When the Kansas reporter called Helen Dukas, then eighty-four years old, she said that she preferred not to discuss the matter, adding only, "I think the whole thing is repugnant."

People who knew Nathan, like Jamie Sayen and his mother, Thomas Bucky and Robert Schulmann, would in the future be surprised to learn that Einstein's surly executor had, for the better part of thirty years, carried on a friendly correspondence with the man who took Einstein's brain. They would find it, at first blush, puzzling that a man so uncompromising on every other point concerning the memory of the late genius, could have lost track of such a precious part of the genius. And it's difficult to know what, if anything, Einstein's secretary knew of Nathan's relationship with Thomas Harvey, or how she might have explained it. Dukas simply adored Nathan, and she considered him an honourable, unselfish man who tirelessly—some would say maniacally—chased down enough leads and amassed enough material to triple the size of the estate's literary archives.

Even after the estate lost its legal battle against the Princeton University Press in 1980, the case having dragged through the courts for two years, Nathan continued covert operations to keep sensitive material from public eyes. The octogenarian trustees knew it would soon be out of their hands, shipped to the Hebrew University in Israel, as Einstein had wished.

Every week, Nathan took the train from New York to Princeton to rummage through the files in Einstein's former study. "We know for a fact that Nathan destroyed things he considered too personal to reveal," Schulmann said. They knew because Stachel oversaw the creation of microfilm duplicates of the Einstein papers, so that the editing work could continue in the United States after the archive had moved to Israel. But they knew, too, because on Wednesdays, when it was safe, Schulmann made his own secret trips to Mercer Street, with a portable photocopier in tow. "We know Nathan destroyed things because there were photocopies made of six documents for which we know of no original," Schulmann said. Among the contested items was a letter that Einstein sent to Elsa after she had made an unpleasant observation concerning her husband-to-be. "She had told him that he smelled bad," Schulmann said. "Einstein wrote 'I'll start taking more baths.' And in the translation, you see Einstein's self-deprecating humour. He signed off, 'a kiss from an appetizing distance.' Nathan destroyed the original of this . . . which really in my opinion was just foolish."

Nathan's obsessive fear of bad press might have been enough to keep his temper in check when dealing with Harvey, Schulmann suspects. "He was probably a little bit scared that Harvey was off the wall and he was especially, deadly afraid of publicity he didn't control." Or perhaps Nathan worried that making an enemy of Harvey might force the pathologist to defend himself and reveal that his meandering analysis of the brain had continued all these years with the blessing of Einstein's executor. Whatever the reason, despite six years of no contact between them, Nathan still mustered a courteous, if formal, tone on December 30, 1981.

Dear Dr. Harvey:

You may be surprised hearing from me after such a long time during which there has been no correspondence.

A friend sent me recently an article which was published in the *Kansas City Times* on October 27, 1981, under the title "Einstein's Brain Still a Convoluted Puzzle."

I learned from the article that you had moved to a small town in Missouri in which you practice medicine apparently as a general physician. I am sure that this is a very satisfying and fulfilling task after the many years which you devoted to laboratory work.

I am writing you to renew our acquaintance of almost 27 years, but also to inquire about your present intention to describe and publish your work on Einstein's brain. So far as I know, you have never published the results of the research you did over a number of years. As we expressed to you a number of times, we are of the opinion that the results of your work should be made accessible to a larger audience, particularly to physicians and scientists who are interested in that problem. A short line from you would be most welcome.

I hope that all has been well with your family and yourself and wish you would be kind enough to remember me to your wife whose acquaintance I was so happy to make when you were living in New Jersey.

With kind regards,
Otto Nathan
Executor and Trustee

☼

For the third time in three years, Einstein dropped anchor in California in 1933. He came with Elsa, thirty pieces of luggage and, this time, his violin. He packed for a long trip, as though he was certain he would never again set foot in Europe. While lecturing on the cosmos at Caltech, so engrossed in his own work that he failed to notice an earthquake that sent the rest of the campus scrambling, the Nazis seized power in Germany. Hitler abolished civil liberties and rival political parties. He threw his opponents into concentration camps and ordered the Gestapo to hunt down his enemies—of which the outspoken Einstein was most certainly one. Countless times the physicist had railed against Germany's fascist policies and Hitler in particular, who Einstein described as "living off the empty stomach of Germany," predicting that the Führer would lose his grip on the nation once economic conditions improved.

To the Nazis, Einstein was a dangerous Jew, dining with kings and queens and telling the world of Nazi atrocities. They did their utmost to discredit him. They burned his books, coerced colleagues to trash his theories, ransacked his summer home at Caputh and his apartment in Berlin. Einstein's stepdaughter and her husband rescued many of his papers from Berlin with the help of the French embassy. Other documents they destroyed themselves to keep them from falling to the Nazis, who had offered a reward for Einstein's capture.

The Nobel laureate meanwhile left California for New York and sailed to the Belgian coast, where he renounced his German citizenship. Over the next few months of exile he travelled from Belgium to England, to visit Oxford scientists and Winston Churchill, and then to Switzerland, to see, for the last time, his younger son Eduard at the psychiatric hospital. In October 1933, with Elsa, his assistant Walther Mayer and Helen Dukas, Einstein sailed back to the United States as a refugee.

✷

A big truck lurched to a stop in front of the Institute for Advanced Study in February 1982. A squad of Israeli soldiers stood guard while workers rolled out a parade of filing cabinets stuffed full of the letters and manuscripts that had passed through Einstein's hands during his seventy-six years. Dukas and Nathan spoke on the telephone. She told him that the archives were at last en route to Jerusalem, their final destination. Nathan told her that he had signed the papers with Hebrew University, completing the transfer. Their work was done.

A few days later, Helen Dukas collapsed. On February 10, at the age of eighty-five, she died. The grief-stricken Nathan told the press it was as though "Einstein died a second death when she died." In many respects it was true. Dukas had been with Einstein longer than either of his wives. She had been his housekeeper, cook, secretary, nurse, public relations handler extraordinaire and so many things besides. Only Dukas could tell with a single glance when Einstein might have penned a particular letter, or pinpoint a corresponding reply, or recite the names and professional titles of the scholars in Einstein's circle. She herself had been an archive. Nathan returned to the frame house on Mercer Street after Dukas's funeral, just as he had the day Einstein died, mourning his partner and the end of an era.

Less than two weeks later, he wrote to Harvey yet again.

Dear Dr. Harvey:

I assume that my letter of December 30, 1981, reached you safely. I am sending this note to say that I would much appreciate the courtesy of a reply from you.

I hope I shall hear from you at an early date.

With many regards,
Otto Nathan

P.S. For your convenience, I attach a copy of my previous letter.

Two months passed before Harvey replied. In a letter dated April 8, 1982, he wrote neatly in longhand on plain paper. He wisely stroked Nathan's ego for a job well done and offered as many details of the town of Weston as he did of Einstein's brain.

Dear Dr. Nathan

It is always a pleasure to hear from you. My apologies for the tardy response. You have been doing a fine job of enabling Einstein's great legacy to reach more people. He did have a real friend in you.

Yes, I am now doing general practice in a town of 1,500 that is 30 miles northwest of Kansas City. It's a town with some colorful history and is peopled with some fine folk. It is on the east side (the Missouri side) of the Missouri River. . . .

Until recently my work on Einstein's brain was at a standstill. But Dr. Hartwig Kuhlenbeck of Philadelphia finds he now has time to work on some aspects of it. So also does Dr. Howard Matzke of the University of Kansas. Kuhlenbeck has lately completed a four-volume work on comparative neuroanatomy. Matzke, also a neuroanatomist, has this year given up the chairmanship of Kan. Univ.'s Dept. of Anatomy. The result is that thanks to them, I am being stimulated to resume activity in that line. Between the three of us, some material should soon be published.

The good woman you knew as my wife in New Jersey now lives in Sarasota, Fla with my two daughters. It will please her very much to know you asked to be remembered to her.

I do hope you are in good health.

My very best wishes,
Thomas S. Harvey

Hartwig Kuhlenbeck had in fact sent Harvey a letter. According to Harvey, Kuhlenbeck said that since his neuroanatomy texts had been published he was finally going to examine the slides from Einstein's brain. Harvey, on the other hand, knew that Howard Matzke was never going to study the brain. But Matzke was the one who could tell him about the researcher in California who had called asking for tissue samples. Marian Diamond had said that she knew Matzke, and Harvey had made a mental note to contact the retired anatomy chairman to ask him about her. "I thought I better find out about her," Harvey said. He wasn't about to send pieces of Einstein's brain off to just anyone.

☼

Nathan sat troubled in his apartment that spring of 1982, Harvey's handwritten reply in his hands. Once again the pathologist was telling him, assuring him, that something would soon be published. Nathan no longer believed it. He doubted everything Harvey now told him and everything Harvey had told him in the past. It drove him to distraction to think he had entrusted Einstein's brain to the wrong man. He dictated his thoughts to an assistant, perhaps pausing and rethinking phrases, careful to be diplomatic but firm.

> Dear Dr. Harvey:
>
> I was glad to receive your letter of April 8, 1982 after not having heard from you for quite some time. I was interested in the description of the town in which you are living now and its history. Unless I am mistaken I read somewhere that you are the only physician in town. I assume that it must be a great satisfaction to serve the people who otherwise would not have recourse to immediate medical help.
>
> I was also interested in what you told me about recent work by two physicians on the anatomy of human brains. Unless I misinterpreted the last paragraph in your letter you intend to co-operate with the two experts mentioned in your letter on some new

research on Einstein's brain. I should much appreciate it if you would keep me informed of developments from time to time. In a few days it will be 27 years since Einstein's death. Since his brain was removed for research purposes I believe we owe the scientific world and also laymen some—positive or negative—results of the research on Einstein's brain.

With kind regards,
Otto Nathan
Executor

Nathan's appeal to Harvey's sense of social responsibility marked a drastic departure from the silence he had once demanded. He was saying, positive *or* negative, he wanted the results made public. In 1955, Nathan had written to a journalist that "publication on the results of the brain analysis will be made, if at all, only through medical and scientific journals." But in three decades, time had changed the vantage point of the observer. Dukas was dead. Nathan's work was otherwise complete. Only the matter of the brain remained unresolved. He wanted it laid to rest. He imagined the bulk of it had been distributed to scientists and that unseemly bits of pickled tissue were all that remained, stashed in cardboard. What could he do? To whom could he appeal? He was no doubt thinking then that his original instincts had been correct. Hans Albert had made a mistake, as he had once told Jamie Sayen, in allowing Harvey to keep the brain. Could he try to force Harvey to turn it over? How would Harvey react? How well did he really know the man, this doctor-cum-scientist, with whom he had exchanged polite letters and visits for three decades?

Recently, Nathan had read a familiar name in the newspaper accounts . . . Harry Zimmerman . . . the scientist involved in the squabble between Princeton and New York. He must have recalled how Montefiore had waited for the brain that never arrived. Nothing had gone smoothly from the moment Einstein's brain was lifted from his head. Now Zimmerman, Nathan thought, was the only other person he knew who might be able to tell him something about the credentials, the credibility, of Thomas Harvey.

Five days after Nathan had dictated his reply to Weston, Missouri, four days after the twenty-seventh anniversary of Einstein's death, the executor dictated a letter bound for the Bronx.

Dear Dr. Zimmerman,

Shortly after the death of Albert Einstein, who passed away 27 years ago last week, I had the pleasure of meeting you in regard to the analysis of Albert Einstein's brain. You may recall that the brain was removed from the body by Dr. Thomas S. Harvey who, at the time, was the pathologist of Princeton Hospital where Einstein died and where the autopsy was performed.

We have been in touch with Dr. Harvey on and off for 27 years. We were anxious that the results of the research undertaken by him, in co-operation with some other pathologists, be made public. Since it had become known that the brain had been removed to be scientifically examined, we have felt that the scientific world as well as the public at large had a right to be advised of the results of the research.

We never succeeded in getting any results from Dr. Harvey. He left Princeton Hospital to establish his own laboratory somewhere in New Jersey. We were in touch with him then and even had some personal meetings. Each time he promised that he would do the necessary upon completion of the research. He then moved to the Midwest and did not reply to our letters.

When we learned lately through a newspaper report that he had now established himself as a private physician in a little town in Missouri, we wrote him again. Copies of our letters are attached.

He finally replied to our second letter on April 8, 1982. A copy of his reply is also attached.

I am addressing myself to you, first, to find out whether Dr. Harvey has ever been in touch with you except in the first few weeks after Einstein's death. He advised us many years ago that the research undertaken had not produced any tangible indications of why Einstein's mind was so superior to that of most other people.

We should also appreciate learning from you whether the two scientists mentioned in Dr. Harvey's letter (Kuhlenbeck and Matzke) are known to you and whether you assume that the co-operation between Dr. Harvey and these two gentlemen promises to produce worthwhile results.

As the Executor of Einstein's Last Will, I am much concerned about the final disposition of Einstein's brain. A few years ago Dr. Harvey was visited by a gentleman from New Jersey who published the results of his visit in a New Jersey [magazine]. He mentioned in his article that, during the interview, Dr. Harvey opened a drawer of his desk and produced a paper carton saying that it contained part of Einstein's brain. I wonder whether the brain, or what is left of it, should be deposited in a special laboratory or in a Medical school or whether it should be cremated as Einstein's body was.

I realize that I have raised some questions in this letter which may be difficult for you to reply to. In any event, there is no need for you to reply to me by letter if that is inconvenient. I should be happy to meet you at any place, on a day and time convenient to both of us, to obtain from you the advice which I am seeking and which you possibly can give me.

With kind regards,
Otto Nathan
Executor

They might have met somewhere in New York: two white heads from another corner of the century, bent toward each other, speaking in polite, earnest tones. Zimmerman would have been eighty years old in 1982, working well past retirement at the Montefiore lab. Perhaps Nathan went to see him there, as he had in April 1955, to demand that he and his hospital say nothing further to the press about Einstein's brain. Or perhaps, with the eighty-nine-year-old Nathan being in frail health, Zimmerman paid a visit to the apartment off Washington Square. There is no record of their exchange.

Around that time, Zimmerman had been taking calls from reporters who must have dug up his name from their archives. He told the Kansas paper that although the methods at the time had been limited, he had finished his investigation of Einstein's brain—or what he had of it—in the 1950s. He said nothing about scientists having had little clue as to what they might actually study. But he did say that "Einstein didn't want anyone talking about his genius cells or nonsense like that. I don't think Dr. Harvey will ever write a report. He doesn't have enough information." He might have winded Nathan if he told him the same thing, if he explained that Harvey was no brain expert, that he had phoned Montefiore the very day of the autopsy looking for help, and that Princeton Hospital was in no way equipped to handle the project. "I feel sort of sorry for Harvey," Zimmerman would later tell a reporter. "He was one of my brightest students. He was always seeking support over the years to help him write a book about Einstein's brain. . . . He, himself, doesn't really know enough about neuropathology to write about it." The executor might have shuddered, too, to hear details of the conversation Zimmerman had once had with Einstein himself. As Zimmerman would recall near the end of his own life, Einstein had agreed that scientists could examine his post-mortem brain but that nothing should ever be written about it. And then, perhaps to bolster his own conclusions, Zimmerman might have expressed his opinion that science would never be able to find the key to genius in a preserved brain. Zimmerman would later tell the press that not even the Albert Einstein College of Medicine would be interested in housing the brain of its namesake. It would represent nothing more than an object of fascination—the very thing Einstein would have detested. Chances are Zimmerman told Nathan that the best fate for Einstein's brain was that it be cremated like the physicist's body, incinerated to a cloud of fine, dark particles rising on a gust of wind and carried away.

☼

A local Weston kid once stopped to play tour guide to a reporter: "You heard of Einstein? You know, that real smart guy? . . . You know where his brain is? . . . Right in there." And then from his bicycle the boy nodded toward the clinic's white building on Main Street, opposite the Weston Historical Museum.

Most locals assumed that was where Harvey kept it. They didn't realize that the genius brain was divided among three jars and four wooden boxes. Harvey never stored all of it at the clinic, only the unsectioned tissue. He thought it safer to keep it in different places. The numbered pieces stayed home, in nearby Leavenworth, where he and Raye had moved in 1982. It was there, in 1984, that he first reached into the jars to fish out four pieces of Einstein's cerebral cortex for Marian Diamond.

Howard Matzke had had only glowing things to say about Diamond —she was an innovative scientist conducting leading-edge work. He would have sent slides—he still had the four boxes—but Diamond said she wanted pieces of the gross material. Out dripped #7 from Einstein's right frontal lobe, #196 and #197 from the left frontal lobe, #226 from the left parietal lobe and #48a of the right parietal lobe. From his homemade maps, Harvey figured these blocks were the closest pieces to correspond with Brodmann's areas #9 and #39, in which Diamond wanted to count glial cells. The project sounded interesting. He had read something about glial cells somewhere in one of the medical journals he received every month. Exactly what, he couldn't recall. He collected so many books and magazines that he had had to rent out a building in Weston to house it all. He had created a little library for himself, and he disappeared into it in rare moments to read. On his homemade sketch, he shaded the brain blocks he would send, writing "Diamond" beside each blackened area. Then he headed to the kitchen to find something to put them in. The fourth time Einstein went to California, he would be going by post.

✳

Marian Diamond strode down to the mailroom in the Life Sciences building on a spring morning in 1983 after Jerry, the mailroom manager, had called to tell her that a parcel had arrived.

"And there it was—it wasn't special delivery or anything. Four little chunks in a mayonnaise jar and one of Harvey's sketches to indicate that the pieces were from the right areas."

She held the jar up to the light and examined the brain bits, tan-coloured, embedded in translucent shells of nitrocellulose, each the size of a sugar cube, floating in alcohol.

"Jerry, guess what I've got in here," she said mischievously.

Jerry, an old-school sort who never felt entirely comfortable around a woman scientist, raised his eyes and shot Diamond an apprehensive look that said, "Oh boy, what's she going to say now?"

"I've got some of Einstein's brain," she declared.

"Oh, come on Marian . . ." he groaned.

"I do," she said emphatically. She didn't wait to see if he believed her. She wanted to get back to the lab and take a closer look. After three years of phone calls, letters and anticipation, Diamond could at last hold pieces of the century's great scientific brain in her hands—like a musician playing Mozart's piano, a hockey player gripping Gretzky's stick. She had easily coiled her fingers around the cortical girth of literally hundreds of other brains, but nothing compared to the emotional rush of touching this one. It left the otherwise eloquent Diamond speechless. "You think to yourself, this is the organ that changed the way we viewed the universe," she said. "There were no words to describe the feeling, no adjectives." When she tried, she said she felt "special and thrilled and excited" that working with Einstein's brain "was a whole body experience."

But Arnold Scheibel's excitement waned quickly once he saw the brain pieces in their celloidin cases and glumly realized the limitations he faced. He could never acquaint himself with the fine architecture of Einstein's neurons. He could not count dendrites or the telltale buds and twigs along their branches. He could not, in short, do the Golgi.

Nearly 130 years after it was developed, the capricious Golgi staining method retains its mystery. The recipe's uncanny ability to highlight the major features of a single brain cell has defied chemical explanation ever since its accidental creation in a nineteenth-century Italian lab. Camillo Golgi, a doctor treating "incurables" in a small town outside Milan, used to brew dyes to study brain tissue in a candle-lit, makeshift lab of the hospital kitchen. One night, after he had hardened bits of brain with potassium dichromate and ammonia, he accidentally knocked a tissue block into a dish of silver nitrate solution. When he retrieved it several weeks later and slid it beneath his microscope, he was astonished. The metallic mixture had somehow stained certain nerve cells in such detail that, for the first time, a scientist could see a neuron in all its complex structure: whole cell bodies, thorny dendrite branches, long, thin axon trunks. They appeared in dramatic black on an amber background, only one brain cell of every hundred or so actually absorbing the stain. The method might have seemed unreliable for its haphazard highlighting pattern, but that was a blessing in disguise. If the stain stuck to every one, all of the cells would be obscured in blinding, uniform blackness.

Golgi published his revolutionary report in 1873, calling this new process "*la reazione nera*," the black reaction. Spanish scientist Santiago

Raman y Cajal mastered the mysterious Golgi stain to show that brain cells are individual, unconnected thinking units. The gap between their branches was afterward dubbed "synapse," for the Greek word meaning "clasp." This boosted the arguments of people like Paul Broca, Oskar Vogt and Korbinian Brodmann, who felt that the discovery of separate cells supported the idea that different parts of the brain controlled different functions. In 1906, Golgi and Cajal shared the Nobel Prize for their work.

Throughout Scheibel's career studying the relationship between cognitive function and the fine anatomy of the brain, the Golgi had been his slide rule. But with Einstein's brain "soaked in a bucket of formalin and each piece . . . encased and absolutely hard like plastic, Golgi would not work." In his lab at the University of California, "the question with Einstein's brain was whether there was anything we could quantify to see if it could be distinguished from other brains," he said. But anything else, he felt, would have been "sort of a measure of desperation." At least his wife could forge ahead with her plans. Celloidin is an ideal preservative for counting glia.

For Diamond it was simply a matter of cutting careful slide samples from Einstein's brain pieces and soaking them in the right puddle of dye. She and her lab technicians had considerable practice achieving the proper stain to distinguish neurons from glia. As she would eventually describe in her report, she and her graduate students had carved sugar-cube-sized chunks of cortex from the frontal and parietal lobes of the eleven men, aged forty-seven to eighty, who had died at the veteran's hospital in Martinez. They had cut deep, dipping just below the grey rind. Then they had frozen three-quarters of the forty-four blocks and embedded the others in celloidin—which turned out to both contrast and mimic the way Harvey had preserved his remarkable specimen. Now, Diamond would do the counting herself in corresponding samples of Einstein's brain, shaved no thicker than six one-thousandths of a millimetre. A cocktail of Kluver-Barrera dyes (another century-old staining recipe that came after the Golgi) outlined neurons and glia in a breathtaking spectrum of blues, fuchsia, violets and vermilion. The slides eventually resembled the tie-dyed patterns once ubiquitous at Berkeley. She picked for comparison the most vividly stained tissue samples from each region of each of the twelve brains. Then, peering into a microscopic lens that magnified the tissue one hundred times, with an eyepiece that amplified that image another ten times, she began the gruelling count in wafers of Einstein's cortex, recording the bigger neurons in one column and the smaller dots of glia in another. She started counting in the second layer of cells in the cortex and continued down into the white matter. Tedious though the work was, Diamond felt exhilarated. "I was anxious," she said. "I really wanted to know, did he have more glial cells?"

Like Jerry in the mailroom, few people believed Diamond was doing it. In her lab, no one except her technician knew about the task at hand. But one day one of the students called out:

"Hey, what are you doing over there, Dr. Diamond?"

"I'm counting cells in Einstein's brain," she replied.

"Oh yeah, great, see you later."

Over several months consumed with counting, Diamond enlisted the aid of a statistician to crunch the numbers, and when the job was finished, with the most conservative interpretation possible, she concluded that Einstein's brain did indeed seem different. In every area, the late physicist had more glia per neuron than the eleven brains from the veterans' hospital. But only in one particular area was the difference statistically significant. Einstein appeared to have an average of 73 percent more glial cells than the control subjects in Brodmann's left area #39, or section #226 by Harvey's map: the left inferior parietal lobe. The result lent itself handily to the hypothesis that the left parietal lobe was bound to have been one of the busiest areas of the great physicist's brain. People who suffered strokes or damage to their left parietal lobes could lose a wide range of basic functions: everything from a sense of personal movement to the ability to calculate.

Scheibel was impressed when he saw the results. "I think her work was beautiful," he said. Diamond herself worried that they had only one gifted human brain in the control group, but she thought at least it was a start. Husband and wife wrote the report together—carefully. They didn't want to be seen to be over-interpreting or exaggerating the results. They called their examination a "potentially meaningful measure of the functional status of the brain." If glia nourish neurons, and Einstein's brain had more of them per neuron than those eleven brains believed to represent the average man, it "might reflect the enhanced use of this tissue in the expression of his unusual conceptual powers in comparison with control brains."

They did not debate where they would submit their report. Scheibel's friend Carmen Clemente had recently started a new journal called *Experimental Neurology*. Scheibel and Diamond had published several papers in it, and they assumed that a study on Einstein's brain would lend the little publication some cachet. They included Harvey's name as a co-author on their report, crediting him for having supplied the tissue, and sent him a copy. But Diamond and Scheibel heard nothing from him. "We had no response from him about what we did," Diamond said, "whether he approved . . . or anything."

✵

Harvey picked it up at his post office box in Weston. After thirty years of keeping Einstein's brain, thirty years without a single published study on the specimen, anyone might have expected him to have turned cartwheels when he read it. Instead he worried. All this time he had told Nathan and everyone else that Einstein's brain was no different from any other brain. Zimmerman had reached the same conclusion, and lately he had been telling the press as much. Now here was the Berkeley paper. He knew controversy would follow as soon as it hit the press. Critics would come out with claws.

TEN

INHERITANCE

☼

THEY HAD GATHERED at the restaurant in the Hyatt Regency Hotel in New Brunswick, New Jersey, in 1991, a group of surgeons from Princeton and Rutgers universities. They lingered over dinner plates and wine, their attention fixed on a slim, sandy-haired young man with a frothy moustache and a cognac-smooth South African accent. Charles Boyd was a biochemist turned geneticist, a gene-hunter, to be precise, and he was telling them all about his latest hunt.

Genetics had been a hot field since the 1953 discovery of the coiled double-ribbon structure of DNA inside human cells. But not until the 1980s had computer technology given scientists tools powerful enough to unwind it and mine the genetic instructions that make and operate a human being. Now they had the ability to cut DNA in specific places to study its fragments, to search for and identify the individual genes that encode everything from baldness to big toes. They could splice and magnify and clone them. They could analyze the proteins that genes manufacture and begin to understand the way they function. And so, they could also dream. Around the world, scientists were working to map the location of every one of the some 40,000 genes on the human chromosome in a study called the Human Genome Project. Researchers at the time thought the number might prove to be higher than 100,000. Scientists like Boyd saw the chance to locate defective genes that make people susceptible to certain diseases. If you could hunt down a culprit gene, you might be able to find a way to correct it.

Boyd's interest lay in the genes that produce collagen. Nearly all of his twelve-year career to that point had been devoted to the fibrous collagen proteins responsible for so many of the mechanical functions of the human body: keeping the brain in place beneath the skull; creating a membrane that coats every neuron; lining the lungs and gut; forming skin, tendons, cartilage, connective tissue and the filtration system of the kidneys. Collagen also constitutes the matrix on which bone is deposited. Without it, the human skeleton would be nothing more than sand. So you can imagine, Boyd explained, all the diseases that a faulty collagen gene might cause.

At the U.S. National Institutes of Health in Maryland, Boyd had helped clone the first collagen gene in 1980. And in 1984, he had pinpointed the one that produces the coating of brain cells. Now, at Rutgers, he was stalking the genes that produce the type of collagen and elastin fibres that together help form the walls of blood vessels. Like a quilt of squash balls, elastin allows vein and artery walls to expand and recoil with blood flow. Weak linings can tear or give, inflating a segment of artery into an aneurysm. Boyd told the other dinner guests that he felt his best hope for hunting the gene responsible was to find a family with a strong history of aortic aneurysms. Among relatives, genetic mutations are easier to spot since there are fewer differences in the DNA to cloud the vista.

"We really need to find these kinds of families," Boyd said. "I expect I'll find a man first because men get these sorts of aneurysms more than women, and then in his family perhaps there will be siblings who have the condition and so forth."

One of the Princeton surgeons at the table piped up then.

"You know, Boyd," he said, "you should get hold of the Einstein family. Einstein died of an aortic aneurysm, and I think his tissue is still around because his brain was taken."

Boyd was fascinated. He vaguely recalled that Einstein had died in Princeton. But he had never known what killed him, and he'd certainly had no idea the great man's brain was still around.

"Where is it?" he asked the others.

No one seemed to know. But, they said, it had been in the news some time back.

☆

If Einstein's brain had been "lost" for more than two decades, as media reports described its years of obscurity in Harvey's care, then with the publication of Marian Diamond's study the brain was found. As Diamond had predicted, publishing in *Experimental Neurology* had brought her friend's fledgling journal a world of attention. Through the first half of 1985, headlines in both hemispheres reported that researchers might have found "Keys to Genius." That scientists might somehow be able to explain the inexplicable, quantify the immeasurable, was too delicious a story to resist. Radio programmers, television producers and reporters from as far away as Australia called Berkeley for the details. But as surely as the report made its splash, so too did skeptics have their say. Walter Reich, a psychiatrist in Washington, D.C., wrote a commentary in the *New York Times* suggesting that the study said more about the power of genius to captivate and beguile us than it did about the causes of genius itself. He chronicled the failure of earlier attempts to link neuroanatomy with intelligence. "Why," he wondered, "do scientists in one generation after the next continue to try?" Reich wrote that "despite the complexity of genius, the possibility of finding and touching it in one spot is so attractive that some neuroanatomists will probably go on trying to do so, especially if they have access to the brain of a world-famous and universally acknowledged genius. . . . [T]o scientists as well as the rest of us, true genius is something so rare, so striking, so unearthly and beyond our ken, that the person who possesses it may be seen as fundamentally different, as having a brain that is somehow wired in a different way."

Reich, along with many other neuroscientists, felt that the California study was too fraught with practical problems to offer a meaningful clue. The Berkeley group, other scientists pointed out, had no information on the mental state of the eleven veterans to whose brains they had compared Einstein's. Nor had they compared exactly the same slices of brain tissue in the late physicist and the veterans; they had only counted cells in one slice from each block. They had only one genius. Their work was too random to be precise. How could they distinguish glia from neurons? If you had a large glial cell or a small neuron, might they not look the same? Neurons might shrink in celloidin; no one knew by how much. Every brain responds differently to different preservation techniques. Maybe an abundance of glial cells was not a good thing—science did not fully understand them. Glia might take the place of neurons as people age. They could be the equivalent of scar tissue. Idiots have big brains loaded with them. It was too wide a leap to consider rat studies a guide to human subjects. With all these problems, it was no wonder they had to publish in a journal like *Experimental Neurology*, the critics said. The prestigious publications had surely rejected the article outright.

Diamond heard it all. "I wanted to boost Carmen's journal," she said. "So I didn't go to *Science* and I didn't go to *Lancet*, so it was never turned down by any of them, because it was accepted in *Experimental Neurology*. Well, among all the big-name journals, *Experimental Neurology* is just a little journal. So that immediately gave fodder to the critics to say, 'Oooh, that's the only place she could get it published.' But no, that was the only place I wanted it published." She knew that any scientist who raises an unconventional notion opens herself to scrutiny. Especially when the data are "statistically shaky," as she herself once described them. But, she argued, the progress of science depends on the confirmation and refuting of new ideas. Fourteen years passed before the world fully accepted Einstein's universe. Yet studying Einstein's brain brought criticism with an ugly undertone, as though researchers who would spend time on such a specimen were seeking mere fame by association, as though there was something slightly tawdry about it. Dr. Albert Galaburda of Beth Israel Hospital in Massachusetts told the *Boston Globe* that finding anatomical differences between brains was hardly unusual since no two brains are exactly alike. But finding physical markers of giftedness was a different thing altogether. "Everyone wants to know the secret," he was quoted as saying. "But serious workers laugh at this kind of work." Some quite literally did.

Bruce Ransom, the editor of the journal *Glia*, who had found Diamond's rat research "nearly offensive," would recall (with regret) years later that "When I went out around the country in the mid-'80s lecturing about the role of glial cells in the brain, I used to take an article with me from the *San Francisco Chronicle* about how doctors find glial cells made Einstein smart. I had it on a slide that I used to put up first to break the ice, and when people saw that they all just roared with laughter."

Arnold Scheibel recalled how critics had once jeered at Oskar Vogt, mocking his work as little more than glorified phrenology. But, like Vogt, Scheibel made no apologies. "It would have been nice if people had thought it was interesting. You'd like people to say, that's a good try. But I think science is competitive, and there's a great deal of ad hominem." The negative reaction did nothing to sway his sense that finding an anatomical basis of genius was a legitimate scientific quest. "The human race has bootstrapped itself up by the 1 percent of the population that is gifted," he said.

Some years later, Scheibel and Diamond would try to expand their study. They contacted the Moscow Brain Institute, asking for tissue from the preserved brain of Nobel Peace Prize winner Andrei Sakharov. The

father of the Soviet hydrogen bomb turned human-rights activist, whose
life so uncannily mirrored Einstein's, had died in 1989. From the leader
who built up the ideology of the Soviet Union to the dissident who
helped to topple it, the appetite of the Moscow brain analysts had come
full circle. Sakharov's brain was among the last added to the eclectic
collection Oskar Vogt had started sixty-three years earlier with the
brain of Vladimir Lenin. Diamond and Scheibel hoped to compare
Sakharov's to Einstein's. The Cold War had thawed by the time they
sent their request, but the institute turned them down nonetheless.
Some secrets, apparently, were still considered too sensitive to share.

✲

A few weeks after dinner at the Hyatt Regency, Boyd wandered over
to the Rutgers library to dig up papers on the cause of Einstein's
death. He found only one: a 1990 report in the "Surgery" section of
the journal *Gynecology & Obstetrics*. The article recounted how doctors in New York had discovered the telltale bulge in Einstein's aorta
in 1949. It was indeed an aneurysm. Boyd started asking colleagues
for suggestions as to how he might find more information about
Einstein's ruptured aorta, his family's medical history or the whereabouts of his brain.

Of all the families in the world who suffered from the condition,
Boyd found himself compelled to investigate the Einstein case. As a boy
growing up in Capetown, South Africa, following in the footsteps of
his chemist father, Boyd had fancied himself something of a budding
Einstein, a scientist in the making, mixing potions in a makeshift laboratory in the family garage. Perhaps he had reassessed that notion on
the day he engineered an explosion while trying to concoct sulphuric
acid and ran screaming from the shed with his jeans on fire. But the "idea
of him, of Einstein the scientist, was an inspiration to young people
everywhere." It wasn't the brain itself that tantalized Boyd; it was the

prospect of what he could literally grind out of it: the organic strings that contained the code that had built "the smartest man this planet has ever produced." Imagine, he thought, analyzing the genetics of that.

A colleague suggested that if Boyd wanted to know something about Einstein he should contact Boston University and speak to Robert Schulmann. Schulmann had moved to Boston from Princeton, bringing with him the duplicate copy of the Einstein papers, in 1984. Since joining the Einstein project three years earlier, Schulmann had toured the world gathering details and documents about the science and private life of the vaunted physicist. Off the top of his head, he could rhyme off the medical history Boyd was looking for. Einstein's father Hermann died at fifty-five of heart disease in 1902 and never witnessed the mind-bending feats of his first-born. His mother Pauline died of stomach cancer in Albert's study in 1920, two years before her son won the Nobel Prize. His sister Maja suffered a stroke in 1946 and died of pneumonia in the Mercer Street house in 1951. A stroke killed fifty-five-year-old Eduard in 1965, at the Swiss psychiatric hospital where he had been confined and effectively abandoned for more than two decades. Hans Albert died after a heart attack in 1973. Happily for Boyd's purposes, the number of strokes and heart attacks suggested one thing: bad arteries.

Schulmann agreed that Boyd might very well be able to track the brain down and examine it for a genetic predisposition to aneurysms. He also thought it might be possible to compare the DNA results with that of some of the physicist's surviving kin. Schulmann told him that Hans Albert had had two children: a son, Bernard, who now had five children of his own, and a daughter, Evelyn, who had been adopted.

But where exactly is the brain? Boyd asked.

Schulmann said he didn't know for sure. Controversy had surrounded its removal but recently, researchers had published a study on it. Boyd's best bet was to find Dr. Thomas Harvey, the pathologist who took it. Last anyone knew, Harvey was somewhere in the Midwest.

"I remember making some really strange telephone calls to places around Midwestern America saying, 'Oh, excuse me, do you know anyone with Einstein's brain?'"

Eventually, Boyd reached Marian Diamond at Berkeley. She told him that Harvey indeed had the brain. She had contacted him in Weston, Missouri, but said she'd lost touch with him shortly after the study had been published in 1985. At least he had a city, Boyd thought. But by the time he dug up a phone number, Harvey had disappeared again, a silvery-backed salmon slithering into the currents of anonymity, too quick to be caught.

☼

It was the spectre of inherited madness, not flimsy arteries, that haunted Einstein. While he dismissed suggestions that his own genius had come from one parent or the other, the notion that insanity flowed through the veins of generations had distracted him from the days of his first marriage. Mileva's sister had been hospitalized for mental instability, and when Einstein detected peculiar traits in Eduard before he was six he feared that his younger son would suffer the same fate. People said it was Eduard who had inherited his father's spark of brilliance, that he too might have been a genius if his mind had not been tormented so. He read unusually early. He memorized massive passages of text. He was browsing newspapers by the time he entered primary school and had mastered the piano. But Eduard was a delicate, emotional boy, impish and shy, with chronic aches in his head and ears. After a youthful bout of unrequited love, he tumbled into deep depression and then suffered a breakdown from which he never recovered. The doctors ultimately diagnosed him with schizophrenia. But for the intermittent bursts of playing Bach or Handel with fury and tenderness, Eduard spent his days chain-smoking, paunchy and despondent. He worshipped his father, but the sickness made Einstein uncomfortable. Although the physicist spent endless hours with pen in hand, he could not bring himself to write to the son he had not seen since fleeing Europe in 1933. When Eduard wrote, Helen Dukas often hid the letters to spare Einstein the anxiety of reading them.

When Hans Albert wanted to marry in 1925, Einstein waged a passionate protest. He did not so much fear the impact of Mileva's heritage passing to the next generation, but what might pass through the suspect bloodline of the bride-to-be. In *The Private Lives of Albert Einstein*, biographers Highfield and Carter wrote that Einstein believed Frieda Knecht's short stature was evidence of dwarfism, and he believed Knecht's mother, who suffered from an overactive thyroid, was "unbalanced." Einstein's objections echoed the complaints of his own mother, who had cursed her son's love for Mileva Maric. "If it were in my power," Pauline Einstein once wrote, "I would make every possible effort to banish her from our horizon." But just as he had defied his mother and married Mileva, so in turn Hans Albert ignored his. He wed Frieda Knecht in 1927. Einstein dreaded the arrival of their first child. In addition to his fear of the genetic heritage, the physicist felt certain the marriage would inevitably fail and that having children would compound the tragedy. But his first grandson, Bernard, became a favourite. He would himself become a physicist, and it was he who inherited Einstein's beloved violin. Hans Albert and Frieda lost their next two children to illness, after which they adopted the baby Evelyn.

☼

Developments in science had brought the media to Harvey's door in 1978. Then the media had brought the brain to the attention of

science. And in 1985, it was scientists who brought the media back
to Harvey, ensuring that the cycle would continue. Diamond's and
Scheibel's report had put Harvey squarely in the crosshairs of the
press. He had become a reluctant celebrity, dodging and weaving
around their questions. Reporters naturally asked him to explain the
findings of the study for which he had been credited as co-author.
These queries he didn't mind so much. Discussing the science of the
grey matter made him feel, well, more like a scientist. "The belief is
that glial cells nourish neurons and there being more in a given person might indicate that that person has a more active brain," Harvey
would explain. He was careful never to give the impression that he
had actually conducted the research. "Those people in California did
a very fine job," he would say. The uncomfortable silences cropped
up when they went on to ask what other studies had found, what
other analysis was under way or how many researchers had the brain.
"Well," Harvey would begin, "it's in a number of places." He would
answer slowly, with succinct phrases, never volunteering any more
than he was asked. He knew that his responses seemed hopelessly
inadequate: "It was a matter of time elapsing and people didn't know
what had become of it. They wanted explanations and I couldn't really
give them; there was this agreement."

Harvey's vow not to speak to the popular press, made the day
after Einstein's death, had become an end in itself. Nathan was no
longer holding him to the terms of that agreement. In his last letter, Einstein's executor had written: "I believe we owe the scientific
world and also laymen some—positive or negative—results of the
research on Einstein's brain." Harvey said he still thought his answers
might be offensive to the Einstein family, but exactly whom he thought
he would be offending is unclear. Hans Albert and Eduard had both
passed away, and Harvey had no contact with the physicist's grandchildren. He knew only that he had once been sworn to secrecy.
Perhaps because of his Quaker beliefs, perhaps because it was a handy

excuse to avoid having to explain the decades of disappointment and dithering, Harvey had no trouble keeping up his end of the arrangement. If Nathan had since changed his mind, Harvey presumed it was only because the executor had grown impatient.

When Gina Maranto, a reporter from *Discover* magazine, called and visited Weston in the spring of 1985, determined to get to the bottom of the story, Harvey told her that it was Princeton Hospital that had requested the brain be taken. Hospital officials meanwhile told Maranto that the brain had been given to Harvey alone. "I meant to say it was normal hospital procedure, you know, because there was permission for an autopsy," Harvey would later say. When Maranto asked Harvey for the names of other researchers who had studied Einstein's brain, he was unusually blunt. "One thing I don't do," he was quoted as saying: "I don't pass out the names of people so they can be badgered by reporters. It's up to them to make the move. If they wanted to publish their results in *Discover* magazine, they would."

Maranto telephoned Harry Zimmerman in New York for information, but he slammed the phone down. Frustrated at every turn, she at last called Otto Nathan. At ninety-two, Nathan was frail and too weak to talk on the phone. He asked Maranto to put her questions in writing and send him a letter. But before hanging up, Nathan repeated what he had said to the *Kansas City Times* four years earlier: "Albert Einstein did *not* want his brain studied." Nathan knew that Einstein had never actually been so unequivocal himself. A few months before his death, the physicist had written to biographer Carl Seelig that he liked the idea of donating his body to science, but that he had left no explicit instructions on the matter since he feared it would be "a theatrical gesture." But the executor was shrewd. By suggesting that the brain's removal had violated the last wishes of Albert Einstein, Nathan tried to do publicly what he had been unable to do privately: pressure Thomas Harvey to explain himself.

Unlike the *Kansas City Times*, *Discover* is distributed across the continent. It is the kind of magazine that other media look to for story ideas. With that single statement from Nathan, in the popular imagination Harvey became a thief. Though he appeared to be a soft-spoken Dr. Jekyll working in obscurity in Weston, Missouri, the insinuation was that in some earlier lifetime, in a dizzying moment in a morgue, he had morphed into a sticky-fingered Mr. Hyde, the man who stole Einstein's brain. "People said hurtful things about him in the press, but none of it hurt him," said his Weston office manager Cheryl Schimmel. "He just didn't care about things like that. He didn't want to defend himself. What would happen then? Two sides would be blasting each other and it would escalate and go back and forth. He didn't want any of that." Even when Harvey saw Diamond and Scheibel on the *Phil Donahue Show* saying nice things about him on television, the publicity made him grumpy, Schimmel said. Harvey blamed himself. "I should have stressed that she shouldn't speak to the lay press. It was my fault. I never told her that there was an agreement not to talk."

As word spread that Harvey might have pinched the brain without permission, the press called Diamond to ask about the ethical implications of studying what appeared to be a stolen specimen. "All I can say is that he gave us tissue belonging to Einstein's brain and I'm very grateful to him," she told one reporter. "It's not that he gave up on the study. I think he was very wise. He waited until there was a database. And I think that's why we got it because we are among the few people who had a database to compare Einstein's brain with." She added that Harvey had preserved the brain very well: "It's as good now as the moment after Einstein died."

When she thought about the debacle years later, Harvey's story would always remind Diamond of the diener who had traded brains for cigarettes in Martinez, California. "We were told that the patients had willed their bodies to science, and it was like Einstein's brain," she said. "We go by what Harvey said—if he said he had permission, you have to accept that he had permission. I trust people until it's proven otherwise . . . you have to . . . We all got a bad rap in this thing."

✧

Late-winter storms brewed off the eastern seaboard in the early days of March 1985 while Otto Nathan squinted through spectacles in his New York apartment. Maranto had put her questions in writing as he had requested. Chances are he did not read her letter right away. She had included something else in the envelope: an article, four pages long. His old heart might have raced when he read its title: "On the Brain of a Scientist: Albert Einstein." He had waited three decades to see words like those on a page, to savour what he hoped would be its conclusions. Did he see the irony? The popular media that had once plagued him for information about the research into Einstein's brain was now supplying him with the first report. You can imagine him there alone, beneath the soaring ceilings of his dim flat, his pint-sized frame hunched in a chair, his head crooked beneath a reading lamp, poring anxiously over the text, puzzling, then rereading. What did it all mean? What did an aged economist know of neuronal glial ratios and parietal lobes? It would have been nice to have had an expert explain it all. It would have been nice if Harvey had sent him the article—or, better yet, delivered it in person so he could ask him, after thirty long years, what the blazes it amounted to. It looked as though these scientists in California had discovered something of merit. But that didn't jibe with what Harvey had always told him. Harvey had said that the brain was probably no different from any other. Meanwhile, these scientists had found something remarkable in just four tiny pieces of it. What, he wondered, would their findings have been if they had had the whole brain to study?

Nathan turned to the reporter's letter. Maranto could not possibly have predicted the weight her questions carried at that moment:
"Is it the case that Dr. Einstein's son, Hans Albert, granted the brain
to the Princeton Hospital, Dr. Harvey himself, or science in general?
Was there an agreement entered into at that time regarding publication or a prohibition of publication? . . . What has your opinion been
regarding the dissection and analysis of Dr. Einstein's brain, whether
the results are published or not, and what is it now? If an agreement
was made initially to let the brain be studied, do you now regret that
decision? By what authority does Dr. Harvey keep the brain?"

Nathan phoned Maranto the day the blizzards hit. He scrawled the
date on top of her letter: "replied by phone March 5, 1985." Maranto
was out—visiting Weston. Nathan left a message.

When she returned his call, Maranto heard the sound of fumbling
and a long silence before the elderly Nathan whispered a hello. He
thanked her for the letter and the copy of Diamond's paper and confessed that he had not understood much of it.

"Unfortunately," he said, "after careful consideration and after
discussing the matter with other [unspecified] people I have decided
I cannot answer your questions . . . thank you for your kindness."

Nathan had been given an opportunity to tell his side of the story,
to vent. But instead he had chosen to keep his feelings, his profound
disappointment, to himself. Perhaps he realized that if he agreed to
an interview he would have to share his own knowledge of the peregrinations of Einstein's brain.

When the May issue of *Discover* carried Maranto's article two
months later, Nathan learned that his silence had done nothing to
cast a softer light on events. "If science with its present knowledge
can learn nothing from Einstein's brain, then why is the organ being
preserved? Whose idea was it, anyway, to study the brain, and why
did it wind up with the obscure Thomas Harvey, an M.D. with no
particular reputation as a neuroscientist?" Maranto wrote. "Why had

Nathan and Hans Albert given their permission to a study of the brain if they knew Einstein was very much against it?" Her conclusion: "What lingers is a sense that the whole affair has become a tawdry botch . . . one wishes heartily Einstein's brain had gone the way of his body, scattered to the winds of the universe, just another blip in the realm of space-time."

A full-page colour photograph accompanied the article: Einstein's brain posed in all its sectioned splendour; a close-up shot of the two jars so clear that you can almost read the numbered labels stuck to its putty-coloured pieces like a cerebral puzzle. Nathan wanted it back. Little had slipped past his scrutinizing eye as executor and trustee of the Einstein estate. So fastidious had he been with every detail that only then, three decades after Einstein's death, was Princeton University Press at last planning to publish the first volume of the physicist's collected papers. The brain, meanwhile, stood apart as a singular matter of unfinished business. It would live on after him in the black hole of the future.

Nathan felt his age with every breath. He had begun to put his own affairs in order and was preparing to transfer his meticulous files on the Einstein estate to the Hebrew University in Jerusalem. He might have turned his attention then to less onerous matters, spent his days visiting old friends or penning his memoirs. But it's funny what a man thinks of when time is running out, and stranger still to whom he turns at the end. For Harry Zimmerman, it would be a young doctor on the night shift. For Otto Nathan, it turned out to be a young reporter in Missouri.

Jeff Truesdell, three years out of journalism school, had profiled Harvey in the *Columbia Daily Tribune* in March 1985. He sent Nathan a copy of the article after it was published, along with a short note sending his regards. After Maranto's article came out in *Discover*, Nathan dictated a letter to Truesdell on May 30, 1985. In it, he offered information that he had never before shared with anyone.

Dear Mr. Truesdell:

It suddenly came to my attention that, because of an error which would be hard to explain, I have never yet acknowledged the receipt of your letter of March 28, 1985 and the two copies of your article. . . .

Since you may want to use the contents of the article some time in another place I hope I may call attention to a few unimportant mistakes in your article:

In one of the first few paragraphs you said that Harvey "received a nod from Einstein's family to study the brain in the interest of science." This is in a way correct but not complete. Einstein died around 1:15 a.m. on April 18, 1955. The autopsy was made in the morning of that day some time after 9 a.m. Einstein was cremated between 4 and 5 p.m. The few people who attended the cremation went back to Einstein's house. Around 6 p.m. Dr. Thomas Harvey came to the house and was received by Einstein's eldest son—and me. Dr. Harvey came to ask permission to study the brain. He was told it was too late since the body had already been cremated. Dr. Harvey replied that when he performed the autopsy in the morning he had taken the brain out since he intended to ask for permission to do the research. Under the circumstances, Einstein's son and I did not refuse permission. I knew that Einstein had once said he would leave orders that his body be given for a school for medical research if he were convinced that a great deal of fuss would never be made of it. So we thought that since the brain had been separated from the body—however without permission, that we should not refuse the study of the brain, particularly since we insisted that there should be as little publicity about it as at all possible and that the results of the study should be published in a scientific magazine and not in a public newspaper. . . . I do not know who is responsible for the publicity that has started . . . and that spread in many newspapers . . .

I hope these few remarks may be useful . . . warm regards

Otto Nathan

The peculiar thing about Nathan's recollection is that Harvey appears more conscientious in his version than he ever described himself as being. Harvey said that he never went to Einstein's house after the autopsy; he only wished that he had, because, he said, "it would have been the right thing to do." When Harvey heard how Nathan described the events in 1985, he said, "I never went to the house to discuss it with the family . . . I guess that's what he thought he remembered. I don't know how old he was when he died, but as we get older, people remember things differently. They sometimes have false memories of things."

Nathan, being ninety-two at the time, was perhaps suffering "false memories." But perhaps Nathan did it deliberately. People in the future would wonder—as Nathan himself likely wondered—why he had not demanded the organ's return when he was younger and stronger and full of fight. If he portrayed Harvey as someone with motives so sincere that he had come the very evening of Einstein's death to ask permission to keep the brain, the world might understand how the organ had been entrusted to this one man, a man whom Nathan had for years considered an ally, a man who preserved the organ so that science might find in it a dramatic footnote to Einstein's brilliant legacy. Perhaps, as Robert Schulmann speculated, Einstein's executor was considering how history might view him. Nathan would, after all, have the letters to Truesdell and every one he had ever written to Harvey shipped to the Einstein archive after his death.

When Truesdell replied to Nathan in September 1985, suggesting that he might publish a follow-up including Nathan's story about the day Einstein died, the executor changed his mind. "I haven't heard from Dr. Harvey in several years," Nathan wrote. "I am well over ninety-two years and in ill health and I am unable to engage in a struggle with Dr. Harvey. The question of ownership of the brain has never been clarified. I am certain that at the meeting of April 18, 1955, Professor Einstein Jr. and I assumed, although we did not state so, that the brain would return to the family after completion of the research to be done. . . . If you can pick out some sentences [from the previous letter] which would serve your purpose and which could not possibly offend Dr. Harvey, I shall be glad to re-examine your request [for an interview]."

After relentlessly censoring, bullying and suing scholars, biographers and Einstein's own family members to keep them from publicizing—or even reading—anything he thought might tarnish Einstein's golden reputation, Nathan was, in his last days, inexplicably worried about "offending" Thomas Harvey. Was it merely an old-fashioned sense of social decorum? Or was it the fear that if he succumbed to publicly airing his grievances, he would open the door for Harvey to publicly respond?

Truesdell never did publish Nathan's remarks in the Columbia paper. Although the young reporter tried to convince Nathan to grant him a formal interview on the subject, the executor refused. In later years, Truesdell, who would become editor of the *Orlando Weekly* in Florida, could only assume Nathan "was so frustrated, so angered, so tired of the stories that Thomas Harvey was telling" that he chose to unburden himself to "a reporter from the sticks." Small-town papers have small staffs and so, by necessity, short attention spans. Truesdell moved on to other stories, chalking up his exchange with Nathan to a case of "airline intimacy . . . where you sit beside a stranger on a plane that you will never meet again and just talk and talk."

Otto Nathan died on January 27, 1987. Jamie Sayen and the few friends Nathan had made in his political crusades and academic circles held a small memorial in New York City. But Einstein scholars shed few tears. Roger Highfield and Paul Carter wrote that one of them actually remarked "that everybody rejoiced" when Nathan passed away. Harvey never mourned. At the time, he thought Einstein's executor had died long before then.

✿

Geneticist Charles Boyd finally found Harvey early in 1993. A woman in Weston told him that Harvey used to be the doctor in town, but he'd retired, she said, and moved across the river to Lawrence, Kansas. As with his departure from Princeton in 1960, controversy tainted Harvey's departure from medicine.

He had decided, after nearly a decade of doctoring in the small Missouri town, to move to North Carolina, where his eldest son, Thomas, lived with his wife and children, and where Harvey had a lucrative offer to work weekends in the emergency ward of a Wilmington hospital. He planned to run a general practice clinic during the week. Money had never driven Harvey. But among the family he'd accrued—three wives and ten children—his indifference to outstanding patient fees and a tax bill he rarely discussed, the doctor was nearly broke.

Shortly after he arrived in North Carolina, the Kansas state medical board called Harvey back to answer to a complaint filed against him. A nurse there alleged that Harvey had mistreated a patient during one of his weekend stints at a Kansas hospital. The patient had come in with troubled breathing. The nurse thought the patient should have been put on a respirator, but the one-time pathologist preferred to draw blood. Oxygen and carbon dioxide levels came back normal, Harvey said— "the patient walked out of his own accord." The Kansas board eventually cleared him, but the case spooked North Carolina. If the senior-citizen Harvey wanted to continue practising there he would first have to pass the Federal Licensing Examination, the same one students take to graduate medical school.

"Most people who take it usually have about three months to study," Harvey said. "But I had not quite a month." The exam covers all areas of general medicine. Some of the subjects, like obstetrics, Harvey had had no experience in since he'd left Yale in 1941. "I studied hard, but I was still practising during the days," he said.

On December 1, 1989, his age showing in the slope of his shoulders, Harvey plodded in to the massive examination hall of one of the state universities. He slid behind a desk, his white hair standing out in the rows of youthful heads. At seventy-seven, he was easily the oldest person in the room. "It lasted for three days, three hours every morning and then a break for lunch and then three hours in the afternoon." North Carolina required a passing score of seventy-five. But Harvey fell just shy of that with a mark of seventy-two. "I think I did pretty well for only having a month to study," he said. "That score would have been good enough for many other states, but not there." He tried to appeal the board's decision. His old internist friend from Princeton, Lou Fishman, wrote a letter "vouching for him," but it made no difference. If people asked about his retirement he had, of course, a plausible explanation: "I was seventy-seven years old and it was getting to be more work then I wanted to do, and so I closed the practice." Cheryl Schimmel said he moved to back to the Midwest because "Rachelle didn't like [North Carolina] so much. . . . They eventually moved to Lawrence because that's where Rachelle's daughter Ginny lived." Retirement did not slow Thomas Harvey. He split and scattered his energies over a thousand obligations. Einstein's brain was just one of them.

Charles Boyd thought that Harvey sounded gentle on the phone, that he had a dignity to his voice. After a lengthy introduction, he asked Harvey whether he still had Einstein's brain. "He seemed quite concerned that I was some kind of bounty hunter," Boyd said. "He wanted to know in great detail about my background and what I hoped to do with the tissue." Harvey was by necessity careful. Each time the media put his co-ordinates back on the map, the keeper of Einstein's brain fended off bundles of mail, much of it bizarre, just as letters from strangers had once filled mailbags for Einstein himself. All sorts of people—housewives, entrepreneurs, the editor of a farmers' almanac—wrote to Harvey inquiring about the organ. Some wanted a piece of it. Some wanted him to dispose of it. It could be hard to make out "the serious scientists" from the pack. But a geneticist, well, that was different. Harvey had a special respect for the field, having scouted through chromosomes himself back in Freehold, New Jersey. Perhaps, he thought, this would turn out as he had suspected from the beginning: *someday we will learn things from this brain that we can't now* ... Perhaps a geneticist would see with his cutting-edge tools things that earlier scientists had missed.

Boyd's name never appeared on the schematics of Einstein's brain, though #47 in the rear area of the right frontal lobe was at some point labelled "mystery block," and a missing piece from the left lobe has a question mark written on it. Harvey might have slipped on his labelling duties, but this time he would not leave a scientist in suspense. A few days after Boyd phoned, he scooped two chunks from the jars and called a courier.

Albert Einstein at age twenty-six in 1905, the year the great physicist published his four revolutionary scientific papers, including his first on relativity. (*Bettman / CORBIS*)

The house at
112 Mercer Street
where Albert
Einstein lived for
twenty-two years
in Princeton.

Albert Einstein reviewing correspondence in his Princeton study in 1940 with the help of his devoted assistant Helen Dukas, who, along with Otto Nathan, became a trustee of the Einstein estate. (*Lucien Aigner/CORBIS*)

Albert Einstein in a New York lounge in December 1930 surrounded by a crush of reporters, as he typically was. (*Bettmann / CORBIS*)

Thomas Harvey on the steps of Princeton Hospital telling reporters the cause of Albert Einstein's death after performing the autopsy on April 18, 1955. (*Bettman / CORBIS*)

Otto Nathan (foreground), executor and co-trustee of Einstein's estate, and Einstein's eldest son, Hans Albert Einstein, leave Princeton Hospital on April 19, 1955, after learning that Einstein's brain has been removed for study. (*Bettman/CORBIS*)

Expecting to receive Einstein's brain for examination, neuropathologist Harry Zimmerman posed for a news photographer on April 19, 1955, at Montefiore Hospital where he was director of laboratories. (*Bettman/CORBIS*)

Thomas Harvey working in his lab at Princeton Hospital ca. 1955. (*courtesy of Robert Harvey*)

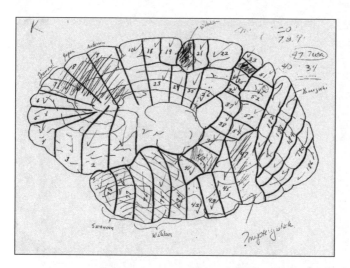

One of several brain maps Thomas Harvey drew of Einstein's brain to keep track of the 240 pieces he carved for microscopic slides to be made. This view represents the right hemisphere of the late physicist's brain. The shaded pieces were sent to scientists to be studied. (*courtesy of Elliot Krauss*)

Two of the three jars containing the sectioned pieces of Einstein's cerebral cortex.

Thomas Harvey at age eighty-two in Lawrence, Kansas, examining slides of Einstein's brain tissue from one of the boxes prepared at the University of Pennsylvania in 1955. (*courtesy of Thomas Harvey*)

McMaster University neuropsychologist Sandra Witelson, outfitted in an Einstein T-shirt (right), and her assistant Debra Kigar posing for Thomas Harvey in Hamilton, Ontario, in January 1996. (*courtesy of Thomas Harvey*)

Debra Kigar, assistant to Sandra Witelson, takes stock of the pieces of Einstein's brain that were spread across a lab table as Thomas Harvey looked on in January 1996. (*courtesy of Thomas Harvey*)

Elliot Krauss, the current chief pathologist at Princeton Medical Center and guardian of Einstein's brain, in February 2000 standing by the steel table where Einstein was autopsied in 1955. The morgue, now used for storage, remains virtually unchanged since Einstein's death.

Thomas Harvey at age eighty-seven standing on the deck of his home in Titusville, New Jersey, in February 2000.

ELEVEN

A DEEP, DARK
SECRET

✵

THIRTY-EIGHT YEARS after Einstein died, two quarter-sized pieces of his famous brain returned to New Jersey. Harvey sent them to Rutgers in the airtight, prepaid Federal Express package Charles Boyd had arranged. "It was a remarkable thing," Boyd said, "that something so ordinary, an everyday bit of life like a FedEx package, should contain this most remarkable thing." Parcel in hand, he happily announced the tissue's arrival to the staff and graduate students. They watched in silent anticipation as he cracked the seal. "We were all amazed," he said. "Here was this brain that had defined our perception of the universe. We felt awestruck." One of the students transferred the cerebral chunks into a small jar, labelled it "Big Al's Brain" and popped it in the fridge. Boyd thought the moniker a tad irreverent and made a mental note to peel the sticker off later. First, he wanted to call Robert Schulmann in Boston, tell him the news and get contact numbers so that he could ask Einstein's surviving family to donate their DNA to the gene hunt.

Schulmann was glad to hear that the brain samples had arrived. He told Boyd that Hans Albert's son Bernard was a physicist living in Switzerland, and that Bernard's eldest son, Thomas, was a doctor in Santa Monica. Then Schulmann gave Boyd the phone number for Hans Albert's daughter, Evelyn. This was a curious thing to relay, considering that the official line on Einstein's granddaughter was that she had been adopted. Her blood would have been useless in a genetic analysis. But Schulmann had made a deal with Boyd; the German history professor had his own question to ask of Einstein's brain.

☆

Since his last visit to Switzerland, the land of Einstein's school days and famed streaks of brilliance, Schulmann had returned to Boston with a tantalizing mystery and few clues to solve it. The trip had begun like so many of his journeys down the paths of the Nobel laureate's life. He dug into archives, consulted local newspaper clippings and rifled through correspondence in various literary estates in search of anything that bore the signature or name of Albert Einstein. As with most of these fishing expeditions, which he'd been making twice a year since joining the Einstein papers project in 1981, Schulmann paid a visit to the prominent Swiss physicist Res Jost and his wife Hilda. Jost had spent a decade alongside Einstein at the Institute for Advanced Study in Princeton, leaving just weeks before the genius died in 1955. He had since joined the European arm of the project and raised 900,000 Swiss francs to support it.

On that particular 1985 trip, the Josts hosted a dinner party at their house in a Zurich suburb and, as Schulmann put it, "Einstein was in the air that evening." Among the guests was the seventy-three-year-old sculptor Gina Zangger, "a tough old lady who would sometimes dribble things out." Her late father Heinrich Zangger had been a professor of forensic medicine at the University of Zurich and a confidant of Einstein's. It was her father, along with the physicist's close friend Michele Besso, who had taken care of Mileva and the boys after the Einsteins' marriage broke up. Gina Zangger still had all sorts of Einstein treasures in her possession, and connections to others who did as well. One day she had come to Schulmann bearing documents from the Swiss patent office that contained typically irreverent notes scrawled by Einstein in the margin. That night at the Josts', Schulmann sensed that Zangger was bursting to share another tidbit. "The psychology of these things is that they want to tell you, but they don't want to tell you." Eventually, Zangger told.

Her story started innocuously. She mentioned that she happened to be a graduate of the same Swiss boarding school that Evelyn Einstein had attended in the 1950s. Then she said that she also happened to be a close friend of the school principal's wife. From her, Zangger said she had learned something about the girl Evelyn that few other people on the planet knew. Zangger said that Evelyn's mother, Frieda Knecht, had told school officials that her adopted daughter was actually the illegitimate daughter of Albert Einstein himself. Knecht apparently said "that she and Hans Albert were raising the girl as a favour to the physicist." Schulmann had never heard such a thing. Yet no one at the table reacted, not even Jost. Schulmann tried to press Zangger for further details, "but she let it drop, and that was all, to tantalize."

It was hardly an outlandish proposition. Einstein's penchant for sleeping with women other than his wives was a badly kept secret. The physicist openly argued that marriage was incompatible with human nature. "Slavery in a cultural garment," he called it, "the unsuccessful attempt to make something lasting out of an incident."

"We knew he had affairs," Schulmann said, "but this, we didn't know."

✵

From the time Einstein was young and handsome, his hair a sea of black waves, he flirted like a Frenchman and scribbled love notes sweet enough to rot teeth. "So crazy with desire," he once wrote to Mileva in 1900, "While thinking of his Dollie / His pillow catches fire." But the fire spread well beyond the marriage bed. The scientist was known to wink and smirk and become completely enraptured at the mere sight of a girl kneading bread. Janos Plesch,

Einstein's doctor and friend in Berlin, was so well aware of the young physicist's "not too discriminating" carnal pursuits that he wondered whether syphilis ultimately killed him. Plesch's son Peter told Einstein biographers Roger Highfield and Paul Carter that "Einstein loved women, and the commoner and sweatier and smellier they were, the better he liked them."

When relativity exploded into a worldwide sensation and rocketed Einstein to fame, there were ample opportunities to indulge. Women hovered in his orbit. They sent flowers with little cards, proposing marriage or romantic interludes. They pounced on him at social functions, pressing him to explain his relativity theories while they basked in the glow of his stardom. "His fame was sexy," said Hilda Jost. "Women were running after him, and he was a very pleasant, gentle man with also a very soft voice." Einstein scholars suspect that the scientist preferred affairs without complications; physics was his only true love. As he wrote in a farewell letter to a younger woman in 1924, he "had to seek in the stars what he was denied on earth." Yet many details surfaced after his death to suggest that Einstein rarely denied himself. His live-in maid at Caputh described a parade of women to the German summer house from 1927 to 1932; scheming blondes and wealthy widows with exotic names like Margerette Lebach and Estella Katzenellenbogen who came bearing pastries and chocolates for Mrs. Einstein and a robust appetite for her husband. Either they whisked him away for the day in their fancy cars, or his second wife, Elsa, whose mournful tears ultimately dried in resignation, buzzed off to Berlin.

The indiscretions continued in America. Thomas Bucky, who visited Einstein while the scientist vacationed in the Adirondacks at Saranac Lake, recounted similar summertime trysts. "He had a few girlfriends, but they were all ugly," Bucky said. "They stayed at the house in the summers. Einstein would take them out on the lake. They were a little bit younger than he was. When I met them I would just

go the other way." Some have speculated that Einstein's compassion-
ate nature drew him to less attractive women. But the late Alice Kahler,
the wife of a German historian who arrived in Princeton before the
war, told Einstein biographer Denis Brian in 1989 that Einstein simply loved women—period. "He was infatuated by the ladies. But he
once said to me, 'The whole thing lasts just ten minutes and then it's
all over.' . . . He wrote on my most precious photo of him and myself
that he regretted I wouldn't sleep with him. I have it in my sleeping
room. I would have been interested in him if I hadn't had a husband."

Right up into his seventies, the genius waxed romantic. Gillett
Griffin, a Princeton art historian who came to know the physicist
shortly before his death, once worked as a curator in the university
library with the woman who called herself Einstein's girlfriend.
Joanna Fanta and Einstein knew each other from Europe, and
Griffin would sometimes accompany her when she visited the little
house on Mercer Street. "Einstein would also call at her home and
read her Goethe and Thomas Mann," Griffin said. "She said that
he wrote her love poems but those are still sealed . . . he also left
her a manuscript, a pencil copy of his work on the unified field
theory—he left them to her so she would never be penniless."
Fanta eventually sold them.

☼

Einstein would have been sixty-two when Evelyn was born in 1941.
Schulmann had no reason to suppose the years had doused the old
professor's libido. Certainly it was possible to father a child at that
age, but was the rumour true? Some time after the dinner party,
Schulmann had a chance to speak alone with Res Jost. When he asked
him what he knew about Gina Zangger's story, the elderly physicist
nodded. He had heard this before, he said. What's more, Jost added
another shadowy detail to the story. Evelyn, he told Schulmann, was
the product of an affair Einstein had with a dancer in New York in

1940. Schulmann tried again to get to the bottom of the story: "But he would never tell me where the information had come from."

When Schulmann attended another dinner at the Josts' house on that same trip, Evelyn's name crept up again in conversation. It was a remarkable coincidence. Other members of the Einstein family had hardly mentioned her in all the years the American scholars had been struggling to piece together the physicist's life. They assumed the family had effectively disowned Evelyn because she was adopted, "an outsider of the bloodline," as Schulmann phrased it. Now her name had popped up twice in one trip.

They had been discussing the famous love letters between Einstein and his first wife, Mileva, the ones Hans Albert's wife Frieda Knecht had planned to publish in 1958 until Otto Nathan sued. Schulmann mentioned that the Einstein scholars had no idea what the letters contained, or whether they had even survived—considering how tenacious Nathan was in hunting down and disposing of sensitive correspondence. Hilda Jost told him that the letters had indeed survived. During the 1979 celebrations of the one-hundredth anniversary of Einstein's birth, Mrs. Jost said she had bumped into Aude Einstein, the wife of Hans Albert's grandson Bernard. "Aude told me the letters were very beautiful, that the family had them," Hilda Jost said. Schulmann said he wanted dearly to track them down; they would add invaluable details to understanding the early years of Einstein's life. Another of the Josts' guests suggested that he visit Evelyn Einstein in California—surely she would know something of the long-lost love letters her parents had once fought to publish. It struck Schulmann then that if he met Evelyn, he might also learn more about her mysterious adoption: "I let the intrigue carry me."

※

A short drive north of Berkeley, where four bits of her grandfather's brain floated in a university lab, Evelyn Einstein lived in a high-rise

apartment that security staff guarded round the clock. She was glad
for the protection: her last name was a magnet for prying strangers;
her dangerous job demanded caution. In 1986, Evelyn was nearing
the halfway mark of a twenty-year stint deprogramming members
of cults. Only in California it seems—with a Moonies indoctrina-
tion camp nestled up the highway in the Napa Valley and an endless
stream of starry-eyed youth flowing west—could you build a career
wringing the minds of the brainwashed. Evelyn considered herself
well trained for the task. She had spent six years in the controlling
environment of the High Alpine Daughters Institute. At the isolated
Swiss boarding school, letters home were read and censored; a Jew,
she said, was forced to take Christian religion classes; and instructors told you how to think and what to think. Her mother had finally
yanked her out at age seventeen. But Evelyn was home only two
months before her mother dropped dead of an aortic aneurysm, the
same arterial beast that killed her grandfather. For a while, Evelyn
lived with her father, Hans Albert, and his new wife, Elizabeth Roboz,
a Hungarian Jew and neurochemist. Freed from the authoritarian
chains of the Daughters Institute, Evelyn grew her hair, cultivated
a fiery independence and dabbled in the leftist passions of the
"Bezerkeley" student protests. She was arrested once, in San Francisco,
along with throngs of young activists that the police had firehosed
down the stairs of city hall, where Senator Joseph McCarthy's House
Committee on Un-American Activities had been meeting.

Evelyn was never close to her older brother Bernard or to her
father's second wife. After her father died in 1973, she did indeed
feel that the rest of the Einstein family had effectively disowned her.
Contrary to what strangers and even acquaintances thought, being
an Einstein provided her no life of luxury. Sure, she'd used the Einstein
connection to her advantage a few times—to land boyfriends in school
and, later, to meet her favourite comedian. Otherwise, she thought,
"the name was a liability." People assumed she was clever as hell, or
snobby and rich. Meanwhile, she bounced from job to job trying to

support herself. She had graduated from Berkeley with a master's degree in comparative literature of the Middle Ages, but she eventually became a child-care worker, a conservation officer, a cop and an anthropology research assistant. When her brief marriage to an anthropologist broke up around the time her father died, Evelyn found herself penniless. Her grandfather's estate sucked in millions in royalties and licensing fees and "I was basically living on the streets for two years," she said, "destitute." Desperate, she once telephoned Otto Nathan, thinking he could arrange to send her a little estate money to get her on her feet, or, perhaps, to go back to school. But she did it reluctantly. She knew her parents felt that Nathan had insinuated himself into her grandfather's affairs, exploiting the gulf between a father and his sons to cement his own role as confidant and caretaker. "He hated the family. He wanted to be the closest one to my grandfather . . . in many ways, I think he was in control of my grandfather," she said. "But I swallowed my pride and I did try to contact Nathan and tell him I was living on the streets. He sneered at me, he almost sounded gleeful. . . . He said he couldn't help."

Evelyn tended to deal with adversity with the same dry humour that distinguished Einstein. She could be brash, funny and rarely diluted her opinions. World-wise and world-weary, Evelyn was suspicious when Schulmann called her that January in 1986. She agreed to meet him—but only in a public place. "I didn't know who this creature was," she said.

Schulmann did not notice a resemblance to Einstein when he first laid eyes on Evelyn. She was not the smallest of women. She had a pale complexion, thick brown hair to her shoulders, a broad forehead and a quick tongue. Only later did he ponder the similarities of their noses, both slightly fleshy and rounded at the tip. But Schulmann had no intention of raising her parentage at that first meeting; he wasn't sure he would bring it up at all.

Evelyn's concerns that Schulmann might be an Einstein fanatic tricking her into meeting evaporated in the first minutes they spent together. "We really clicked, hit it off right away," she said. Evelyn ended up telling Schulmann about the dysfunctions of her family and why she thought her relatives had essentially ignored her for the last dozen years. Schulmann, meanwhile, described the struggle of the Einstein papers project and how, largely because Helen Dukas and Otto Nathan thwarted their efforts, not a single volume of her grandfather's collected works had been published thirty-one years after his death. Then he broached the subject of her grandparents' lost correspondence.

Did she know what happened to the love letters?

No, Evelyn said, she didn't. She had only the handwritten introduction her mother had written for the book that Nathan had blocked.

Evelyn went home after their meeting and pulled her mother's manuscript from its plastic folder. Inside, behind the pages, she discovered that she did indeed have some portions of the letters. They were extracts, intended for publication, from correspondence between Einstein and his elder son, and, more important, exchanges between Einstein and Mileva from the turn of the century.

That afternoon in Evelyn's apartment, Schulmann glimpsed the details Nathan had battled so hard to stifle. With these extracts testifying to the historical value the letters held, the Einstein scholars opened negotiations with the Hebrew University and Einstein's family to find and publish the originals. "Nathan had not yet passed away, but Hebrew University was more amenable to compromise," Schulmann said. "If he turned out to be a wife beater or a homosexual, so be it . . . they did not hold these kinds of things so dear to their little hearts as Nathan did."

For years, Hans Albert's widow, Elizabeth Roboz, had claimed to have no knowledge of the letters, though they had remained in the Berkeley house where she lived for decades. Only a few years earlier she had transferred them to a safety deposit box. On April 18, 1986, exactly thirty-one years after Einstein's death, the correspondence was removed from the vault; Enstein scholar John Stachel and a representative from the Hebrew University stood by to photocopy every sheet of paper.

The letters opened new windows for researchers to explore Einstein's scientific thinking. For the first time the world learned the possible contributions Mileva Maric made to the theory of relativity. So too did they contain evidence that Einstein was human after all. Just as he wrote passionate poetry, so could he also pen brutal notes. Setting out the conditions under which he would continue their marriage, Einstein wrote to Mileva in July 1914: "A. You will see to it (1) that my clothes and linen are kept in order, (2) that I am served three regular meals a day in my room. B. You will renounce all personal relations with me, except when these are required to keep up social appearances. . . . You will expect no affection from me. . . . You must leave my bedroom or study at once without protesting when I ask you to."

Most intriguing were the letters indicating that Einstein had in fact sired at least one illegitimate daughter. The year before they married, in 1902, Mileva gave birth to a baby girl named Lieserl. Einstein was over the moon in the beginning. He wanted to know everything about the baby; he even asked for a sketch of her. But a few days later all mention of the newborn disappears from their correspondence. No one knows for sure whether Lieserl was put up for adoption or if she died shortly after birth. But later evidence suggests that Einstein suspected she was alive, even after he moved to America. In 1935, when Lieserl would have been thirty-three, the physicist hired a private detective to investigate a woman in Europe claiming to be his long-lost daughter. Schulmann would eventually discover letters from the detective filtered through Helen Dukas, meaning that Einstein's assistant—and no doubt Nathan—had been well aware of the secret. The correspondence appeared in the first volume of *The Collected Papers of Albert Einstein* in 1987. Death spared Nathan, by mere months, from witnessing its publication. But the executor might have taken some comfort. Although the headlines announced that "Letters shatter Einstein's popular image, show him as unkind husband," the public did not revel in the physicist's dirty laundry. In New York, Christie's expected the personal letters, which were included in a collection of Einstein papers, to fetch more than $2 million on the auction block in 1996. But the bidding brought in less than half that. Chris Coover, a manuscript expert at Christie's auction house, told a reporter at the time, "People are unwilling to look at Einstein's darker side. The harsher side is less appealing. They prefer that he conform to the genial genius."

☼

Schulmann and Evelyn became friends during the archival quest. The historian came to admire her directness, her irreverent humour. In turn, Evelyn grew to trust him. When Schulmann visited California again later in 1986, Evelyn decided to share something that had haunted her much of her life. "I figured there were things he should know, about my suspicions. I wanted him to know where I was coming from," she said. They met at a restaurant in a strip plaza not far from her apartment. "I remember that morning. I hadn't told many people and I was as nervous as a cat."

"Robert," she said, "there is something I want to tell you, but I'm so nervous about it.'"

"What is it?" he replied. "That you might be Albert Einstein's daughter?"

"He said that he'd heard that all over Europe. I was completely slayed by that. Here was someone totally unconnected to me who knew about this, this thing that I had only ever mentioned to three people I was closest to because I didn't want people to think I was nuts. To me it was a deep, deep dark secret."

Before her parents had shipped her off to the High Alpine Daughters Institute at age eleven, they had told Evelyn she was adopted. She was too young to remember details of the conversation. What she did recall were moments when her mother told her that "there were things I needed to know when I was old enough to understand." But her mother never lived to tell her these things.

Suspicion took root. The first troubling detail was her 1941 birth date. Her parents had been living in South Carolina then, German refugees who had escaped Europe in 1938, bringing with them only what they could carry. Until Hans Albert found work as a researcher with the Department of Agriculture, Einstein gave them money to support themselves. "Why would a couple counting pennies, who had lost everything, decide to adopt?" Evelyn thought. Of course, she realized that her parents had lost two children. Diphtheria killed their six-year-old son Klaus the same year they arrived in America. Then they had lost a baby girl. Born with a defective stomach valve, the infant had effectively starved to death. Her mother had remained in hospital for about six months afterward suffering some "womanly problems," Evelyn said, which would have been around the time of her adoption. "According to my brother [Bernard] he had to take care of me a lot when I was a baby," she said, "so I wondered why would they be adopting when my mother was sick in the hospital. This was not the time to be adopting."

Her adoption papers arrived in the mail after her father died. Documents from his estate indicated that she had been born in Chicago. "Why," she wondered, "would they go all the way from South Carolina to Chicago to adopt?" Eventually, she plucked up the courage to contact the woman named as her biological mother on her birth certificate, to ask for details, "and I thought perhaps she might want to meet." Instead, the woman offered "a weird response," she said, "a totally inappropriate comment about the fact that I couldn't black-mail her. That was the end of the conversation."

Comments once made casually began to acquire an eerie significance as the years passed. She thought back to an incident just before her baptism at the Swiss boarding school. Students who were baptized and confirmed were excused from further religion classes and Evelyn decided that a Gentile christening would be her ticket out. She picked her favourite teacher to be her godfather, and her mother flew from America to attend. On the day-long train ride from Zurich to the school in the St. Moritz valley, Frieda Knecht coincidentally sat next to the man who happened to be the father of Evelyn's favourite teacher. "They had nine hours on the train to talk, and I think my mother must have told him the whole story, because when he arrived he was very excited about something," Evelyn said. She learned later that her mother had apparently described how horribly upset Einstein was when he learned that Mileva had baptized Hans Albert and Eduard as orthodox Christians, in keeping with the Maric family's religion. Knecht had then gone on to say, "It was a good thing Einstein was dead or he'd be seeing another of his children baptized."

The possibility that she was Einstein's daughter comforted Evelyn. It beamed new light on the frosty relationship she had with her father, and a childhood often drenched in unspeakable sadness. All those years in the Alps she'd felt abandoned, like her Uncle Eduard. Many times she wondered why her father kept his distance from her. If they argued, and she ran from the house, "he locked it up as tight as a drum to keep me out . . . the worst thing he did was let me know that I was a nobody. He gave me this constant impression that he didn't want me and I remember thinking, 'If these people adopted me—didn't they want me?' I think he must have felt a resentment at the fact that his father told him to take care of his bastard child, and that he couldn't say no to his father, to Albert Einstein, even if this was the man who had mistreated his mamma . . . how do you say no to the most famous scientist in the world?"

Einstein never took a special interest in her that she remembered. His letters to her were always warm and fun-loving. But when she wrote to him, Helen Dukas often confiscated the letters and replied on her grandfather's behalf. Dukas offered excuses —"which I found offensive"—along the lines that Einstein was too busy to respond, as if a young girl had no place in the physicist's geodesic world. Yet given the mysterious fate of his first illegitimate daughter, Lieserl, Evelyn suspected that Einstein might have felt a particular responsibility to her after she was born: "Maybe he wanted to make sure that this female offspring stayed in the family."

Not until Charles Boyd called about Einstein's brain in 1992 did Schulmann harbour hope of finding a definitive answer. "Jesus," he thought. "This is perfect." The stories passed on were not enough to convince him. As a historian, he wanted hard proof—if there was any to be had—like the letters that documented Lieserl's birth. But so far, inquiries had yielded little more than rumours. As Evelyn put it, "I've heard my mother was a dancer, a ballerina, an artiste. She might have been Marilyn Monroe!" Now here was an opportunity to unravel the truth like a biological scroll. Schulmann raised the issue with the geneticist immediately. "I said I could help make a deal here and help get him these [DNA] samples if he would be willing to do a paternity test . . . Evelyn was very eager to find out."

With a *Star Trek* brooch often pinned to her lapel and in her pocket a laser-light key chain that can beam UFO images onto any surface from a considerable distance, Evelyn had long been a science buff. When Boyd phoned her, she quizzed him for details of his multigenerational project: "I love that stuff, that sort of scientific investigation. I love the idea of genetics. . . . I was really excited." Boyd was relieved to have her co-operation. None of the other family members returned his calls. Evelyn, meanwhile, readily agreed to see a dermatologist, who "literally punched a circle of flesh and blood" out of her arm and mailed the sample to New Jersey, where Boyd was set to purée half a piece of Einstein's brain.

DNA testing of dead celebrities became such a common quest of the 1990s that scientists would eventually poke through the molecular remains of all sorts of notable personalities—from Beethoven to Butch Cassidy, Jesse James to Thomas Jefferson. The curious hoped to harvest intimate details of historical figures not available in archives, digging into the very essence of their former beings to settle paternity claims and questions about their health. Around the time that Boyd took his shot at reading Einstein's genes, a southern California company came up with a venture, called Stargene, to capitalize on celebrity DNA. It was started by the unorthodox biotechnologist Kary Mullis, who had won the 1993 Nobel Chemistry Prize for inventing the revolutionary method that lets researchers amplify a single gene and within hours make millions of identical copies. Mullis planned to mass-produce the DNA of distinguished people, dead and alive. He would then encase a biological smudge, along with their pictures, on hockey- or baseball-type trading cards. But the estates of Marilyn Monroe, Elvis Presley and many living celebrities turned him down. They worried what secrets might be unearthed from the genetic code. In 1992, one of Stargene's executives told a London newspaper that they wanted to get their hands on Einstein's DNA: "Parts of his brain are said to be preserved . . . if we can ascertain who holds the rights to his name, that would be wonderful."

✻

Be it brain or beetle or even broccoli, the steps for extracting DNA from organic tissue, human or otherwise, are essentially the same. With a teacup, measuring spoon, salt, water, cheesecloth, liquid detergent, meat tenderizer, rubbing alcohol and a clean jar, harvesting genetic material from a banana, for example, takes only minutes. Dice the banana, add it to the teacup with a teaspoon of salt, mash until smooth with one tablespoon of water and all the fruit's cells separate from one another. Strain the mixture through the cheesecloth over a jar and add a teaspoon of detergent to the liquid, stirring carefully so as not to create bubbles. Let it stand five minutes and the soap dissolves the fatty casings around each cell and the shell around the nucleus within. Add a few dashes of meat tenderizer and stir gently, allowing its enzymes to literally cut through the remaining proteins of the membranes like a pair of scissors, leaving the DNA floating and exposed. Tilt the jar to drizzle an equal part of rubbing alcohol down the side and it will form a liquid layer over the surface of the heavier banana juice. The remaining proteins, fats and sugars sink to the bottom. The white stringy clumps of DNA rise to the top, ready to be skimmed.

Boyd experimented with countless home recipes in his Capetown family garage. Now here he was, surrounded by state-of-the art equipment in a university lab, with "Big Al's Brain" awaiting his attention. He cut one of the small tissue blocks in half, dropped it in a blender and ground it to the consistency of apple sauce. Then he followed the banana process, substituting the chemical phenol for salt, detergent and tenderizer and relied on a slow-speed centrifuge to separate the cells. The spinning coaxed the proteins to the surface, which were then siphoned off before alcohol was added to the remaining mixture. Once the alcohol evaporated, only the raw form of Einstein's DNA remained.

DNA is composed of four different chemical units known as nucleotides, strung together in a long chain. They are almost always referred to in their short form, as A, G, T and C. Together they represent the chemical alphabet of humanity. In each gene, the letters are arranged in a unique order, spelling out different chemical messages to design and operate an organism. Magnified, the DNA chain is actually two parallel chains of nucleotides, spiralling around one another in such a way that the A pairs with the T, and G pairs with C. Each coupling is called a base pair, and within each human cell there are three billion of them.

If Boyd were to read anything meaningful in Einstein's DNA, there had to be enough base pairs left to distinguish genes from junk. He tried to tell himself there was no sense stoking high expectations of tissue preserved for thirty-eight years. It was, after all, still a few years before scientists would tease DNA from the bones and teeth of five-thousand-year-old Egyptian mummies and discover that ancient peoples suffered cancers and tuberculosis just as their descendants would. But Boyd had some reason to hope. Earlier research showed that at least twenty thousand legible base pairs could survive in human tissue preserved in celloidin. So the deciphering began.

He stained the DNA with ethidium bromide to run it through the tabletop scanner that scientists call a sequencing machine. "It's like a fluorescent box that has a glass top and an ultraviolet light underneath it," he said. "We put the DNA in gel on top of the glass surface and with the light underneath you can see the stained DNA. It shows up as white bands." But even with his naked eye, Boyd could tell there wasn't enough. The bands were too short, too broken up, just dashes of Morse code. Hundreds of thousands of chemical letters appeared to be missing from Einstein's genetic recipe. "It had completely fragmented, completely denatured, you could only read one hundred or two hundred base pairs." As far as Boyd could tell, the DNA from Einstein's brain looked as though it had been chewed up, partly by the microorganisms that feast on lifeless humans, and partly by the cannibalizing enzymes in its own tissue.

Boyd forwarded the other half piece of Einstein's brain to colleagues at the genetics laboratory at the National Institutes of Health in Maryland. But they got the same results. Only years later, when he happened to meet Tom Harvey in Princeton, did he discover another reason for the DNA's deterioration. On that warm spring day in 1955, when the structure of DNA was but a two-year-old discovery, Harvey had perfused the brain with a fixative at room temperature. "There's no question," Boyd said, "if he'd used cold formaldehyde, there would have been more DNA left to study." As it was, Einstein's genetic code had melted as surely as a chocolate-covered cone in July.

TWELVE

WORKING-CLASS HERO

✵

THOMAS HARVEY ENDED up in apartment 13. He'd lost his medical licence and his third wife, and instead of enjoying tranquillity in his twilight years, relaxed and surrounded by grandchildren, he was sweating for a paycheque.

Soon after they'd moved to Lawrence, Kansas, in 1989, Rachelle had left him. "She just didn't want to live with me any more," Harvey said. He told her he'd still support her financially, and he took a job with the Census Bureau. For fifteen dollars an hour, with his curving spine and a clipboard, he trudged door to door recording the number of occupants per dwelling. A year later, at seventy-eight, he went to work on the extrusion line of a sweltering plastics factory, the E and E Display Group, making notched shelves for greeting card displays. He had to wear safety goggles on top of his spectacles and protective earmuffs over his hearing aid. By the time he reached his eightieth birthday, the once ambitious doctor, civic leader and celebrated son of Ottumwa was a labourer on the late shift, sharing a cheap flat with a university student. Anyone looking for evidence that fate had made Harvey pay for his sins would have found it there in the walk-up adjacent to a gas station, in the aptly numbered apartment, where the penniless pathologist slept on a sofa bed and hid Einstein's brain in the closet.

All through history, in religion and literature and popular culture, the threat of the curse has cropped up to inspire obedience and
fear: to keep Eve from the forbidden fruit, robbers from graves and
precious books and hidden treasures. In the Middle Ages, when the
fervour for saints' corpses reached its height, people believed that
these relics had the power not only to shower believers with bless-
ings but to damn them as well for irreverent behaviour. Stories of
the blasphemous falling mute or suffering violent deaths fuelled
devotion and served the useful purpose of convincing looters to
leave sacred body parts in their shrines. In the nineteenth century,
after archaeologists unearthed the first ancient Egyptian tombs,
Victorian novelists invented the concept of a mummy's curse awaiting those who disturbed the dead. When the English Lord Carnarvon
died of an infected mosquito bite seven weeks after he opened the
burial chamber of Tutankhamen in 1923, the literary motif landed
in newspapers as a chilling explanation for this and other tragedies
linked to the expedition team. Readers lapped it up: a policeman
guarding King Tut's gold funerary mask in San Francisco in the 1970s
apparently tried to claim compensation for the effects of the alleged
hex after suffering a mild stroke. In the case of India's seventeenth-
century Hope Diamond, the public seemed to take some comfort
in the notion that no good could come from possessing so gigantic
a jewel, that somehow owning the lustrous blue gem could lead a
king and queen to the guillotine, the wealthy to bankruptcy and
death. Similarly, the odyssey of Einstein's brain hinted at a deeper
story, a morality tale suggesting that greed—for saintly skeletal
remains, a sparkling stone or scientific glory—is never rewarded.

For those who considered Harvey a thief, guilty of defying the
wishes of a dying man—and not just any man, but a scientist of near-
celestial stature—his downward mobility suggested retribution.
English journalist Jonathan Freedland, who tracked Harvey to
Lawrence in 1994, described his as a life of penance: "He is 82 and

works till 11:30 each night in the kind of place you would not want your grandpa to spend five minutes. The dust lies in blankets in his spectacles."

Harvey dismissed the idea that his actions had somehow jinxed him as utter nonsense. His decision to keep Einstein's brain arose from a legitimate interest in research, he felt; knowledge was a blessing, not a curse. Sure, there were things he wished had turned out differently. He wished there had been more fruitful study of the organ. He wished he had not lost his medical licence. He wished his third marriage had lasted. But Harvey would never have described himself as a tormented soul. He remained unrepentant: "I did the right thing," he said. "I would do it all over again. . . . I had been trained in pathology many years, I knew how to handle it and take care of it. I think the kind of scientist Einstein was, it's what he would have wanted, too."

Yet it was there, living a hard-luck life in Lawrence, a city with the motto "From ashes to immortality," that Harvey spoke for the first time about giving up the brain. When a reporter for the *Wall Street Journal* discovered Harvey and the closeted brain in the prairie city— just as Levy from the *New Jersey Monthly* had found him with the intriguing cider box fourteen years earlier—the elderly pathologist expressed concern about the fate of the organ. While Harvey said he was in good health for a man his age, one could not predict how or when the forces of nature might strike him down. Sounding oddly superstitious, he explained: "I might get hit by a car, or an earthquake, or a tornado." But he'd made no provisions for his special specimen in the event of his death, and who, he wondered, would take it?

✻

Somewhere in California in 1994, one of the most despised scholars on earth mulled over the front page of the May 5 *Wall Street*

Journal. Just where he sat as he read about the uncertain future of Einstein's brain is anyone's guess. Arthur Jensen would never have dared publicize his address. Police had once advised him to move out of his own house; he booked dinner reservations under the alias Mr. Jones. The seventy-one-year-old educational psychologist, a well-groomed man partial to grey flannels and gold-buttoned navy blazers, had been effectively forced into hiding after he'd first advanced his theories on race and intelligence at Berkeley twenty-five years earlier. While scientists like Marian Diamond crafted rat experiments to learn how nurture affects mental ability, Jensen developed tests to determine what limits nature sets on it. Using examinations he touted as culturally unbiased, he tested blacks, Mexican Americans and other minority schoolchildren. His results supported a two-pronged theory of intelligence that the English psychologist Charles Spearman had pioneered in 1904. In Spearman's model, there were two different types of intelligence. One type involved the ability to retain and recall facts, like naming the capital of Italy or the boiling point of water. The second was considered "general intelligence," which allowed a person to think conceptually, to quickly manipulate known facts and solve problems. Jensen found that the higher the level of general intelligence, what Spearman had called "the g factor," the better a person performed in the widest range of tasks. But the psychologist went further. While memory-based intelligence, based on education, was found to be distributed equally among all races, his results indicated that Asians and whites tended to have a higher general intelligence level than Hispanics and blacks. General intelligence, Jensen deduced, was fundamentally an inherited trait—and some ethnic groups did not inherit as much of it as others.

In February 1969, Jensen published his bold conclusions in the *Harvard Educational Review* under the title "How Much Can We Boost IQ and Scholastic Achievement?" The 123-page article implied that throwing money at education programs for poor black children was unlikely to make them smarter.

Jensen's world erupted. Against the backdrop of the civil rights movement, in the liberal bosom of 1960s Berkeley, he was quickly branded a racist. Professors refused to ride the elevator with him. Students hissed when they passed him in the halls, slashed his car tires and spray-painted swastikas on his door. In his breast pocket, he had to carry a personal alarm connected to the police station. The reception was the same when he travelled abroad, where police had to extricate him from baying mobs and field warnings from would-be assassins. "I have been called a Nazi, a fascist, a racist— all the names in the book," he once said. "There have been times when there were so many death threats that I was convinced my number was up." The reaction surprised Jensen, but it didn't deter him. "I rather resented the implication that in psychology the idea of being politically correct should take precedence over being scientifically correct," he said. "To me that idea is abhorrent."

Jensen continued to conduct exhaustive research into the so-called g factor. But he had begun to feel that testing intellectual performance could go only so far in establishing "g" as a biological property with genetic roots. By 1994, he saw the future of understanding general intelligence in the physical analysis of brains. Having just retired from Berkeley, he was about to write a summary of his life's work. The hefty tome, his seventh book, would be published in 1998 under the title *The g Factor*. In it, Jensen would make the case that physical clues to intelligence exist: brain size, the speed at which signals travel between neurons (which research suggested could be as quick as 100 metres per second), the slower rate at which the brain burns sugar while performing a mental task. As well, he felt another biological argument for "g" was that children in gifted classes show an incidence of near-sightedness three to five times greater than children in regular classes.

Jensen was in the midst of gathering such facts when he noticed the article about Einstein's brain. Ironically, just as the opposite-thinking Diamond had done a decade earlier, Jensen couldn't help but ponder what treasures might be buried in its physical structure. Gifted people fascinated him. In part, they challenged his faith in IQ tests as full measures of intellectual ability. As he himself would write, "genius implies exceptional creativity . . . though such exceptional creativity is conspicuously lacking in the vast majority of people who have a high IQ, though it's probably impossible to find any creative geniuses with a low IQ." As a university student in New York in 1955, Jensen had attended a memorial service for Einstein just to hear what people had to say about him. Now here, he thought, was an opportunity to link that brilliance to biology. "The fact that Einstein's brain had been kept, that he was considered the greatest modern-day genius, made it a wonderful source for hypothesis," he said. "You can't really prove anything with it because it's a single brain and it could be that some of the anatomical differences you find might have had nothing to do with his genius. But as a source of hypothesis, it seemed worthwhile to study."

As a Jew and a socialist who fought much of his life to end racism, Einstein had particular empathy for the blacks of America. A rabbi who raised the plight of "Negroes" in the United States with the physicist shortly after he had moved to Princeton wrote, "[T]here came into Einstein's eyes that look of moral indignation I had seen when he had spoken of German cruelty. His reply was brief and emphatic: 'there should be no discrimination. It is wrong.'" The physicist would have been mortified to think that a scholar like Jensen hoped to use his brain to advance theories that only deepened racial divides. Not that Jensen hoped to study Einstein's brain himself. Biology, per se, wasn't his bag. But recently he'd been corresponding with a young neurologist in Alabama who, like himself, braved the prickly politics of researching intelligence.

✾

Each time Britt Anderson went before the university promotions and tenure committee he'd show up neatly combed and pressed to recite the goals of his research and review his latest endeavours to link brain biology with behaviour. Then, like the trains in Tokyo, the question would arrive right on schedule.

So Dr. Anderson, they would ask, where do you publish your work? In the journal *Intelligence*, he'd tell them.

Hmmmm, they'd say, *Intelligence* . . . anywhere else?

Not really, he'd reply.

Then he'd be left stewing in his Sunday best while they made their notes.

Sometimes committee members inquired about his discoveries or what projects loomed on the horizon. But he suspected they were only being polite: he hadn't been promoted in five years, not since joining the University of Alabama in 1991. He understood their reluctance. To advance, a professor looking for a physical explanation as to why some people are born smarter than others is risky business. "For one, studying intelligence just doesn't have the same emotional plea as finding a cure for schizophrenia or figuring out brain tumours," he said. But more than that, Anderson knew that finding a biological basis for intelligence frightened as much as it fascinated: gather enough evidence that people are not created equal and the very foundation of political democracies might crumble. "This country, our constitution, is based on the idea that anyone can be president of the United States," he said. "You don't want to trumpet the idea that mental ability is entrenched." The twentieth century provided enough horrific examples of the impact of doing so. Look for differences between individuals, and the suggestion of differences between races invariably crops up. Next thing you know one ethnic group is tagged as being mentally inferior to another and the road to Auschwitz is paved again.

Anderson fully appreciated the fear of faculties, politicians and the public at large. Most intelligence research was accumulated as a spinoff of other investigations; the government didn't hand out grants for it. After all, this was America, where IQ tests considered culturally skewed ranked blacks below whites in mental agility. And this was Birmingham, the southern city infamous for fighting the black civil rights movement with firehoses and German shepherds. "Studying intelligence is not really a wise career quest for me," he admitted. "If I had other research questions I was interested in or published other places I would be further ahead . . . but I don't think it's controversial to say that there are obviously differences in certain cognitive functions, the way one person can write beautiful poetry and fail trigonometry."

Still, Anderson couldn't help himself. Even back in 1988, while serving three years on a military base in Biloxi, Mississippi, in return for the U.S. Air Force scholarship that put him through medical school at the University of Southern California, he had pursued his unwelcome investigations. Officially, he was treating patients with neurological conditions at the base hospital. Unofficially, he was studying the brains of bright mice. Applying the theory of general intelligence to rodents, he worked from the assumption that mice that scored better in one type of test would score better on all types of tests. He built them a maze with tire spikes and a pile of Froot Loops at the finish line. To measure their response flexibility, he constructed a runway and dropped an obstruction in the middle of it, watching how the mice found their way around it. Then he evaluated their attention to novelty by plunking a rock or a pop can in their midst. After the tests were complete, Anderson killed his thirty mice and weighed their brains. On average, rodents with heavier brains got better scores. He counted their neurons as a possible explanation for greater brain weight and ability but found little correlation. The only thing he knew for sure when he'd finished the experiment was that it wouldn't be his last.

As soon as his stint at Biloxi was up, Anderson set his sights on employment at an academic medical centre, where he could devote more time to basic scientific research. But because of his military commitment he'd never completed a fellowship to boost his credentials after graduation. It hurt his chances of landing a university faculty position. Then the job came up in Birmingham. To graduates in New York and Washington, he said, the University of Alabama was not a first-choice destination; too many people still associated it with the prejudices of the Deep South. But he thought the characterizations unfair. Besides, UAB offered everything he needed: an assistant professorship in behavioural neurology conducting research and seeing patients. By the time he arrived, with his wife and children, he'd refined his mouse studies to analyze the microscopic components of brain cells.

There are easily one hundred subtypes of neurons. Each one is classified in the same way a tree is identified as a pine, or maple or willow—by the architecture of its branches. The dendrite nerve fibres that fire off messages to neighbouring axons also sprout from the surface of each brain cell in distinct patterns. Scientists actually refer to them as "arborizations of the dendritic tree." Chandelier cells look as though they could hang above a dining-room table. Pyramidal cells resemble the figure geometry implies. Anderson had a hunch that the composition of dendrites played a major role in mental ability. In Alzheimer's patients, for example, dendrites shrivelled. But in other cases, patients seemed to have no physical signs of the ailment, and Anderson decided to take a closer look.

Dendrites that shoot out messages for another neuron to fire up seemed to have tiny spines along their lengths, like thorns on the boughs of a rose bush. If each spine represented a point of connection with another cell, Anderson suspected that Alzheimer's patients with no other physical trademarks of the ailment in their brains might have fewer of them. The trick was finding out whether this was true.

You had to be a master of the Golgi method to achieve a stain that would provide that fine level of detail. But Anderson worked hard on his technique. In 1995, he published his results: dwindling dendritic spines were indeed further evidence of Alzheimer's disease.

Arthur Jensen sent a letter to congratulate him. The retired Berkeley professor had first written to him a few years earlier, introducing himself as an advocate of Anderson's brand of research, and the two had spent time together whenever they met at scientific conferences. Along with this particular letter, Jensen included an article he'd clipped from the *Wall Street Journal* the year before. It was a bizarre story about an old pathologist who might not have anywhere to leave Einstein's brain when he died. The pathologist was quoted as saying he still thought the specimen was worth studying, that scientists might one day be able to learn something useful from it. "Jensen thought I might want to try some of my experiments on Einstein's brain," Anderson said.

Having read the Diamond and Scheibel paper on Einstein's brain in 1985, Anderson knew of the organ's preservation. But he'd never thought about tracking it down himself. Now, the more he thought about it, the more seductive the idea became. A project like that could push all the right buttons: science and celebrity and a safe topic. What could be politically incorrect about investigating the source of genius? Very little, he thought. There is a fine balance between tasteful and distasteful research; between the acceptability of hunting a gene for high IQ and the taboo of finding one linked to low IQ.

Anderson thought he might be able to apply his latest research and study the dendritic spines of a brain whose staggering synaptic connections had changed the face of science. It might, he imagined, be his ticket to academic freedom. He remembered the publicity Diamond and Scheibel had received. This could provide a defining moment in his career. He stared for a few moments at the dateline of the article: Lawrence, Kansas. What the heck, he thought. He called directory assistance and jotted down a number and address for Thomas Harvey. Then he wrote a letter of request and mailed it out with his CV.

While Anderson waited for a reply by mail he practised his staining techniques. Tissue from Alzheimer's patients and rodent specimens was hardly adequate preparation when you were expecting to have the brain of Albert Einstein to slide beneath a microscope. He had heard there was a neuropsychologist up in Canada with a fabulous bank of normal brains, a woman by the name of Sandra Witelson at McMaster University in Hamilton, Ontario. It was clear that she considered her specimens precious. When he phoned, Witelson wanted to know his credentials, what and where he had published. She was hesitant, but finally she agreed to send him one sample to see what his stains could produce. "It was waste tissue," he said. "But I stained it with the Golgi method and sent the tissue back to Hamilton." The work impressed Witelson enough that she sent Anderson tissue from three other brains.

After several weeks and no word from Harvey, Anderson decided to call him. "I had a sense he was glad to hear from me, glad of my interest in the brain," he said. Harvey was indeed happier than Anderson could have imagined. At that time, an inquiry from a neuroscientist came as a welcome tonic. Harvey told Anderson that most of the brain had been sectioned into embedded pieces, but otherwise he offered no information on the state of the organ or other attempts to study it.

Anderson couldn't help thinking that if Einstein had died at the end of the century, there was little chance his brain would have ended up in the care of a single man. "It probably reflects a time when the world was less institutional," Anderson thought. Harvey agreed to send pieces to him, and Anderson promised to arrange a prepaid Federal Express kit with a container for the tissue and a form Harvey could sign and send back. "Very good," Harvey said, with the impeccable manners of a bygone generation. "He seemed," Anderson thought, "like a bit of an eccentric guy."

�֍

Strangers rarely understood Thomas Harvey. That was hardly surprising, since he had mystified co-workers and left three wives bewildered. Yet a man nearly a quarter his age, from a distant corner of
the globe, found a kindred spirit in the old doctor who turned out
to be the best roommate he ever had. Ahilleas Maurellis, a swarthy
young man with a ponytail hanging between his broad shoulders,
first met Harvey at a Sunday Quaker meeting in 1992. Harvey had
long since attained the status of an elder, or a "weighty Quaker" as
they called him, and Maurellis was a twenty-four-year-old South
African student on a Fulbright scholarship at the University of
Kansas. "We took a liking to each other in that indefinable way that
two people somehow click," Maurellis said. "I saw something in him
and he saw something in me."

Maurellis knew no one in Lawrence, and he was looking for a friend
among his fellow Quakers. The cotton-haired Harvey, whose curiosity and energy belied his age, charmed him. The dark and husky
Maurellis, meanwhile, working toward his doctorate in physics and
astronomy, fascinated Harvey for reasons the student could not then
possibly understand. They locked frequently in intense discussion.
Harvey was bursting with philosophical questions about how religion fit into the study of the stars and heavens, just as people had once
peppered Einstein with queries about his religious beliefs. "Try and
penetrate with our limited means the secrets of nature and you will
find that, behind all the discernible concatenations, there remains
something subtle, intangible and inexplicable. Veneration for this
force beyond anything that we can comprehend is my religion,"
Einstein had once answered.

But along with Harvey's devout religious views, he also displayed
the same progressive, liberal attitudes that had led him to advocate
birth control back in 1950s Princeton. The Quaker congregation hit

a stalemate over the morality of same-sex marriages one Sunday morning, and Maurellis recalled Harvey listening intently to the debate. Then, after some time, the weighty Quaker got up and said the Friends should allow it. "He said it was simply the right thing to do. This was his compassion speaking, not just empty tolerance. I thought that was really something."

Destiny drew Maurellis and Harvey together in needy times. Maurellis was fed up with temperamental landlords and moving from apartment to apartment. Harvey was tired of living alone. Maurellis described him as heartbroken that his marriage had ended. One night, before they moved in together, Harvey invited Maurellis for supper at the small flat where he was living. He told Maurellis—or Archie, as he would come to call him—that he'd once been a pathologist in Princeton, that he had met Albert Einstein and performed his autopsy and removed his brain—which, he said, he still had. Maurellis was "flabbergasted." But after Harvey told him the whole story, Maurellis became intensely amused. Although he was a physics student, Maurellis would find no *gravitas* in rooming with the jars that contained the brain of the world's greatest physicist. "For me, I find Einstein in his work, which I have been privileged to study," Maurellis said. "He's not there in bits of tissue. He has a far greater presence in the recorded thoughts that came from it."

In the $400 apartment they rented on Massachusetts Street, the two men, from two opposite ends of the world and the century, slipped easily into their lives together. "I never thought of myself as living with an old man," Maurellis said. "He was more like a senior chum." After midnight, when Harvey trundled in off the late shift at the plastics factory, he'd fix himself a gin martini and steak tartare or a couple of fried eggs and poke his nose into Maurellis's room. Invariably, the student would still be working away, studying the atmosphere of Jupiter, preparing to construct a mathematical computer simulation of the planet's ionosphere. Harvey would sit on the bed and the person so many knew as a man of few words would gab late into the night. He never doled out fatherly advice, but if Maurellis wanted a wise word or two, especially on the subject of women, Harvey gladly gave it. Sometimes Maurellis would explain Einstein's theories, and he was surprised to discover that Harvey knew very few personal details about the genius whose brain he kept, "not even that Einstein generally went without socks."

Unfortunately, the mornings sometimes offered a ritual Maurellis could have done without. In keeping with some invisible calendar, the elderly doctor would climb out of bed and shuffle into the kitchen with the three jars containing the pieces of Einstein's brain. "It was certainly a little disturbing some mornings," Maurellis said. "I'd come into the kitchen to make my sandwiches for school and Dr. Harvey would come in with these tissue samples and retop them with grain alcohol. It was a little stomach-turning."

Maurellis never saw Harvey study the brain. As far as he could tell, the matter did not in any way consume him. Yet to the world, the tissue in his closet had become the defining fact of his life, and the media never let him forget it. In a cable universe where the press had firmly entrenched itself as a power broker of democracy, the days when a reporter could easily be turned away were long gone. Writers and film crews flew to Harvey from all ends of the earth. Inevitably, Maurellis was pulled into the dance of preparations. Harvey grumbled in anticipation while he and his roommate rushed about tidying the place, trying to find a home for the medical library that populated the apartment floor in towering stacks and bloated cardboard boxes.

Every week, Harvey's mailbox bulged with the *Journal of Clinical Pathology*, the *Journal of the American Medical Association*, *Science* and the *New England Journal of Medicine*—his personal favourite since the days when he squatted on a cot in the cold air of the Gaylord Sanatorium. Wall-to-wall bookshelves were the first and only thing Harvey contributed to the furnishings of the tiny two-bedroom. On them he arranged his enormous reading collection: texts on Quakerism, travel, history and medical books. But the shelves were not enough, and occasionally the spillover drove Maurellis nuts. "I always had to climb over these piles of unread journals," he said. "Every time people came over we had this mad dash to clean the place up."

Einstein had cared so little for appearances that it frustrated his second wife, Elsa, to no end. She was smitten with the glamour her husband's celebrity provided, accepting dinner invitations he never wanted, feigning a fancier lifestyle than the physicist would permit. When Elsa once urged him repeatedly to dress up for important company, Einstein quipped that if the important visitors were coming to see him, they'd see him as he was; but if they were coming to see his clothes, Elsa could show them to his wardrobe.

Maurellis could tell that Harvey didn't care for the publicity. "It wasn't for him exciting. He felt people were just snooping into his private life, and because he had the brain, it somehow gave them licence to do that," he said. "He was not a celebrity figure, he was very much a scholar." But Harvey was never depicted in that way. Instead, he became the caricature of Kansas. "Like the Wizard of Oz, he should be a big scary mad scientist," British journalist Jonathan Freedland wrote. "But like the Wizard of Oz, he is nothing of the sort. He is a harmless old buffer, with a rasping laugh and a hearing aid. And, saddest of all, like the Wizard, he can't do any tricks."

In his old age, Harvey had no curtain to hide behind. With Otto Nathan long dead, he was free to exercise his own discretion in the interviews he granted. Perhaps he was naïve, "entirely too trusting" his children thought, allowing himself to be exposed and exploited in the spotlight of celebrity. Yet even the basic facts of his story were so strange that supermarket tabloids would have been hard-pressed to invent them: an anonymous octogenarian doctor, in a backwater town in middle America, having the brain of the most ubiquitous

personality on the planet. So powerful was the allure that Kengi
Sugimoto, the Einstein-obsessed Japanese math professor who had
written to Nathan in 1979 for information about the brain, spent
more than twelve years searching for Thomas Harvey. When he finally
did find him, a BBC film crew was there to record the denouement
of his quest in all its darkly comic pathos. Director Kevin Hull caught
on camera the perplexed face of Harry Zimmerman, who mistakenly
tells his visitors that Thomas Harvey had died two years earlier.
Zimmerman also added his own version of events, claiming that it was
he who phoned Harvey the day Einstein died. "I called Tom Harvey
and asked him if he'd be willing to do the autopsy, or should I come
to Princeton and do it, and he said he's willing to do it." Evelyn Einstein,
who thought the crew was assembling a biography of her grandfather,
said after the interview, "I felt a little sorry for this Sugimoto guy. He
was clearly an Einstein fanatic . . . I couldn't help him." From there,
Sugimoto went on to confound Charles Boyd: "This Japanese math
professor wanted to know if there was any link between aneurysms
and intelligence, if you can imagine."

At last, the road led to Kansas. Surrounded by Harvey's cardboard
chaos, Sugimoto—who, as a native of Nagasaki, had every reason to
detest Einstein's atomic achievements—fell to his knees and nearly
dissolved into tears of joy when the pathologist presented his jars.

"If possible, please give me one piece of Einstein's brain," Sugimoto
pleads in the 1994 documentary.

Harvey retrieves from the kitchen a carving knife, a breadboard
and a fork. Then his fingers, unwithered despite his age, fight to pry
the lid from one of the jars: "It's been there so long," he says, "it's practically one piece now."

While Sugimoto holds his breath, Harvey seems surreal on celluloid as he stands at a makeshift desk and slices Einstein's cerebellum
like a Christmas mushroom.

Later, at a Kansas City nightclub, Sugimoto waves his pickled trophy in one hand and a microphone in the other, crooning a heartfelt

Japanese ballad while the Midwestern crowd cheers and sways on the dance floor. His precious relic would eventually find itself stored in a Twinings tea canister on his desk at Kinki University, the centrepiece of his mind-numbing collection of more than one thousand Einstein photographs, Einstein stamps from seventy countries, Einstein T-shirts, books, posters and coins.

When Cheryl Schimmel, Harvey's friend and former office manager from Weston, heard that the elderly doctor had actually handed out a piece of Einstein's brain as a souvenir, she said, "Well, that Japanese professor must have been a very smart man. Dr. Harvey must have seen something in him to give him some of it."

✿

The fallout of media attention was always the same. Letters streamed in from strangers longing for their own pieces of the brain. After the *Wall Street Journal* article in 1994, which quoted Harvey's musings about turning it over to someone else, the Lawrence flat was flooded. In yet another carton, Harvey kept all of them.

Dear Dr. Harvey,

I am an aspiring neuroscientist. . . . I would like you to consider entrusting me with a portion of Einstein's brain. I am fascinated with the genius of the mind, and I would work very hard to understand some of the human brain's immense power. . . . If you should decide to offer me this opportunity, I promise to try to make my future research the foundation of a doctoral dissertation.

Intellectually curious,
D.E.
San Francisco, California

Dear Dr. Harvey,

We are developing an arts and sciences museum here in Enid, Oklahoma. We would love to have a piece of the brain to excite children and promote their study of this man. We will have a life sciences room with skeletons and skulls and preserved animal parts. This would be a very welcome addition to our collection.

Sincerely,
H.G.
Enid, Oklahoma

Dear Dr. Harvey,

As a scientist and retired headmaster of a large secondary school I should be honoured to receive a small section of this vital organ of our greatest scientist and mathematician.

Please advise me whether or not this is possible and how to proceed . . . it would be greatly appreciated.
With good wishes for Christmas and the New Year

Yours sincerely,
M.L.F.
Warwickshire, England

A Boston insurance agent with Prudential Securities phoned before he wrote, perhaps laying the groundwork to write up an unusual policy.

Dear Dr. Harvey,

It was a pleasure speaking with you. You certainly are an interesting man. . . . Just the fact that you are from West Hartford and dated a girl from my hometown of Granby leads me to believe this, not to mention the fact that you also have Einstein's brain in your closet. . . .

Certainly, given the magnitude of Einstein's mental accomplishments, it would seem someone or some institution would take interest. . . . Wouldn't it be a great shame ten years from now to deny neurological pioneers access to what is arguably the greatest mind in modern history? I think so.

While I may not be a scientist myself or have a medical background, I do respect history and the technological advances of mankind enough, to realize that what you possess in your closet is of significant value and should not be dismissed.

If there is any way that I can assist you in your quest, I would be more than willing to help. . . .

Best Regards,
M.N.
Boston, Massachusetts

Letters that offered Harvey money for the brain made him laugh. There had been several over the years, from an anatomical society in Scotland, from a California research foundation, and from a man offering $15,000 for a single piece of it. Maurellis remembered Harvey chuckling after he read them, "as much as to say they'll never get it." His roommate found it a peculiar response from a man who, for the most part, seemed dirt poor. Harvey kept suits from days when people drove Studebakers. If clothes tore, he mended them himself with needle and thread. "He hoarded everything," Maurellis said, "every nut and bolt in a giant tool box in the basement that he'd hauled with him from place to place, and he never wasted any scrap of food that came into the apartment. . . . I got the impression, and it was only my impression, that a lot of money was going out of his account. He seemed to be writing a lot of cheques to his family and children. It seemed he was almost working for them." Harvey was always very proud of the fact that all of his children—even his stepchildren—graduated from university. But his old Princeton friend Lou Fishman said that Harvey "had to borrow money from me to buy the basic necessities of life. . . . He's an over-do-gooder, if you ask me. To this day, he owns very little himself; any benefit he has is a benefit to his children and his grandchildren in need." But then, Maurellis described charity as the defining fact of Harvey's life. Despite "pretty back-breaking shifts," he said, Harvey would be there in force with the other Quakers cooking and serving food in the soup kitchens, building ramps for the disabled at the Quaker meeting house, weeding its garden and repairing windows. He even made Thanksgiving dinner for a group of Russian immigrants who had come to work at the factory. "He was very much a man of the Depression years. He believed in hard work, in community work . . . the desire to contribute is a major facet of his character."

While his frugal, Quaker ways baffled friends, Harvey's work ethic astounded co-workers at the E and E Display Group. His young supervisor, William Katz, who attended the University of Kansas by day and worked the sprawling factory floor by night, was not surprised that a former doctor should find himself on the line: "People's fortunes can change pretty quick," he said. Colleagues knew Harvey had assorted kids to support, and there was talk too of a long-overdue debt to the taxman. He had started out as a temporary labourer on the four-to-midnight shift, packing parts in boxes. But he rocketed up the ranks. The eighty-year-old was soon operating the giant extruder from a platform with no railing, two and a half metres off the ground. The extrusion machine sucks plastic pellets into a hopper, then heats them in a storey-high vessel before squeezing them out as shelving parts. The slight Harvey could gain enough height to reach into the machine only if he climbed atop a wooden box they kept on the platform for him. "It was not an easy job; it required real physical dexterity. Some people thought we were crazy to let him try it, but he wanted to do it," Katz said. "Tom would work overtime all the time if you let him. I'd be surprised if he was making more than ten dollars an hour." His age-defying feats and hours clocked earned Harvey the employee-of-the-month award in October 1992. His plaque still hangs in the company's corridor. "This was a big deal," Katz explained. The old doctor beat out E and E's 250-odd other employees.

The only time Harvey missed work he had very good reasons: a leg broken in two places, five cracked ribs and stitches across his head. He had been cycling to work when a car hit him. "It wasn't my fault and it wasn't the driver's fault," he said. "Are you familiar with the screen shot in basketball?" The vehicle zipped through from behind another car, smacking Harvey clear off his bike: "I put a dent in the hood with my head, but I didn't crack my skull," he explained, "though my scalp was lacerated." Harvey spent three weeks in the hospital. Doctors had to put a steel rod in his left leg. It came out shorter than his right, and he went home with a limp. Four months later he was back on his perch at the factory.

If he was a man of uncertain scientific stature held up to ridicule in the media, he was a legendary star in the lunch-bucket pageantry of the prairie plant. To Chris Cottrell, the master extruder who trained Harvey, the old doctor "was probably the smartest man I ever met in my life." Cottrell was taking a course on electronics at the time, and Harvey seemed to know the textbook material off the top of his head: the boiling temperatures for certain metals, the composition of plastics, the history of conductors and magnetism. In turn, Cottrell helped Harvey at his acreage outside Lawrence, where the machinist repaired the elderly doctor's old toasters and broken coffee pots. Cottrell even managed to straighten out Harvey's mangled bicycle so he could keep riding it to work. Co-workers gushed years later over Harvey's stamina and stoicism, particularly recalling the day he stood silent after his hand was accidentally sucked between the rollers of a machine he was cleaning. "It was stuck in there for a few minutes. Most people would have been screaming in pain but Tom didn't say anything," Katz said. He was rushed to hospital again, his hand swollen like a baseball mitt.

Harvey earned their awe without any of them knowing he was the man who had Einstein's brain. Only after Katz saw an article about him in the university campus newspaper did word get around. His co-workers thought it was a hoot. But even then, it was no more than factory gossip that soon faded into the drone of machines and coffee-break banter.

✤

Weeks after Anderson had forwarded the FedEx kit to Lawrence in the summer of 1995, the brain had still not turned up. Anxious, he phoned to ask about the delay. Harvey told him that he'd already sent it—just not by courier. Anderson felt sick. The thought of Einstein's brain lost in the U.S. Postal Service was too tragic to contemplate. He called down to the main university mailroom.

"Is there anything there with my name on it?" he asked.

The clerk put the receiver down for a moment and rifled through a drawer of unclaimed packages.

"Oh, yes," the clerk said. "There's a brown envelope, but it's addressed to neurosurgery instead of neurology, and the zip code is incorrect."

Anderson barely hung up before he was out the door: "I knew there were these pieces floating around inside and these people had no idea that they'd had Einstein's brain in their office these last days."

He clutched the brown envelope with both hands as he strolled back down University Boulevard, as heady with anticipation as a child the night before Christmas, knowing that as soon as he returned to his lab he'd tear it open and pull out the gift: a chunk of Einstein's motor cortex from the frontal lobe, a block of his temporal lobe believed to be involved in language and two pieces of his cerebellum. Harvey had sent them in plastic Ziploc bags filled partway with alcohol, the way a pet store packs goldfish for the ride home. Anderson unzipped the pouches and held the pieces in his hands. "I felt giddy, I found myself giggling and chuckling for no good reason," he said. "I don't know whether it looks good on me to say it, but there was just something about being in proximity to it." Anderson worried about leaving it there in the lab he shared with other researchers but thought its best disguise would be no disguise at all. He transferred the pieces into a jar, labelled it "Harvey's tissue" and placed it in the same cabinet as all the other specimens. He learned later that one of the lab students who knew its true identity led colleagues on midnight pilgrimages to ogle it.

After the initial thrill had subsided, Anderson was confronted with the same frustrating fact that Arnold Scheibel had faced: the pieces were embedded in celloidin, jeopardizing his plan to analyze the fine architecture of Einstein's dendrites with a Golgi-type stain. It shouldn't have been a complete surprise, he knew. Celloidin was a commonly used preservative in the first half of the century. But in his disappointment, Anderson couldn't help but think that by the 1950s, if scientists wanted to take a good look at a brain they'd have used paraffin wax, or cut a fresh brain in two, dropped half in formalin and frozen the other half for study. "If Harvey had had a more active collaboration with a leading neuroscientist of his day, he may have been aware that this celloidin doesn't allow optimal use of the tissue," he said. To say nothing of the fact that, by that point, the tissue had been imprisoned in its nitrocellulose shell for forty years.

Anderson placed a second distressed call to Lawrence.

"Others tried to do a Golgi stain," Harvey told him, "but they could not do it."

Anderson tried anyway, marinating a small segment of one of the tissue pieces in Golgi's metallic salt-and-silver recipe for several days. But beneath the microscope, instead of the hallmark pale-lemon background with the occasional well-defined neuron outlined in black, there were only blobs, charcoal smudges of brain cells. He could see nothing. "We are technically limited," he said to himself. "Here we've got Einstein's brain and we're wondering what can we do with it."

He reflected on his short career studying intelligence and thought back to his mice and Froot Loop days, when he had counted neurons in rodent brain tissue. The results of the endeavour had not been particularly illuminating, but if he could use an ordinary stain on Einstein's brain that would allow him to count individual cells, he might be able to evaluate their density in the cortex. With that, he could explore the hypothesis that cells packed closer together, like granules in a sugar cube, allow for easier synaptic connections between their axons and dendrites and therefore result in a richer network of communication between neurons. True, it would only be an inference, but it seemed better than giving up. Besides, he wouldn't be the first researcher to make such a deduction. Earlier that year, in May 1995, Sandra Witelson at McMaster University had published a landmark report showing that neurons are packed more densely in the brains of women than in the brains of men in areas involved in language and speech. Looking at the temporal lobes of the post-mortem brains of five women and four men, at a level parallel to the temples just behind the eye, Witelson had found that the women's five thousand brain cells per cubic millimetre of tissue was 11 percent greater than the number in men. While both sexes score equally well in intelligence tests, Witelson's team suggested that this greater density of neurons might explain why young girls tend to read, speak and pick up languages faster than their male counterparts. In other words, density might give girls an advantage in this particular mental ability.

Anderson thought highly of Witelson's work, and he figured if he could complete a neuron count in Einstein's brain, he would at least have a basis for comparison—assuming that he could round up enough brains of similar age to constitute a control group.

Of the four brain pieces, Anderson decided to use the three-centimetre-wide, five-millimetre-thick tissue block Harvey had labelled #9 on the homemade map included in the envelope. Number 9 likely lay directly beneath Einstein's famed hairline, in the bulbous tissue of the frontal lobes that had for decades been considered wasted terrain in the geography of the human brain. If New England railway worker Phineas Gage could still walk, talk, read and write after an 1848 explosion shot a metal rod clear through his left cheek and frontal lobes, scientists of the nineteenth century to those of the post-war lobotomy era assumed that it must be a fairly dispensable brain area. Only later in the twentieth century did researchers attach strong significance to the fact that the accident had transformed Gage from a churchgoing family man into a profane drunken gambler who turned up in the circus as a medical curiosity. By the time Anderson was selecting which piece to study, science regarded the frontal lobes as the source of thoughts, emotions and behaviour, shaping personality and defining humanity. It's where researchers suspect the punch lines of jokes are processed, where creativity, working memory, planning and problem-solving originate. As well, it is said to contain the moral compass of the brain, enabling one to make delicate social judgments. People who suffer damage to the pre-frontal lobe area that sits squarely behind the eyeballs can be sociopathic. The trademark selfishness and insensitivity of teenagers might be explained by research showing that connections among neurons in the frontal lobes might not fully develop until people reach their twenties or even their thirties. But Anderson's wedge of the frontal lobes had been cut from an area known as the motor cortex, which is believed to fire a signal that a mere fraction of a second later commands muscles to contract in various parts of the body: directing vocal organs to articulate words; fingers to curl around the hand of an old friend or the end of an oar, a violin bow, a pipe or pen.

Anderson trimmed his sample of Einstein's motor cortex to a sliver one millimetre square. He soaked it in the violet-blue Nissl stain for thirty minutes, rinsed it with distilled water and soaked it in alcohol to dry up the excess staining solution. From local brain banks he had collected five sample brains of elderly men, and he cut and stained corresponding bits of tissue from their frontal lobes. Then he counted.

Usually the density of neurons decreases as you travel deeper into the six layers of the brain's cortex . But in Einstein's brain, Anderson found the neurons more evenly distributed in each. Yet, overall, Einstein's brain had roughly the same number of neurons as the sample sections in the control subjects, and the cells were roughly the same size. The only discernible difference was that Einstein's cortex appeared to be thinner. If Einstein's thinner cortex had the same number of neurons as the others, did this not suggest that Einstein's cortex might be more densely packed with brain cells than the control group? There were two implications if this was true, Anderson thought. "It could be that forty years in storage had led to an artificial density," he said. "The second and more romantic possibility is that this reflects an innate difference between Einstein and my control brains." The bottom line was that it meant something . . . or it didn't. But in Anderson's estimation, the results, tenuous though they might be, were worth publishing—not only for the sake of what muted light they shed on the mechanics of one of the most admired scientific brains since Isaac Newton, but because they might further his career. Anderson expected his small study on Einstein's brain to turn him into a newsmaker. In turn, publicity would lend credibility to his work, and he could pursue future intelligence research unfettered. He notified the public relations office at the University of Alabama that it should prepare a press release: "I expected I would get a splash." He intended to submit his paper to *Neuroscience Letters*, an Irish-based science journal that promptly publishes short but complete reports related to brain studies. Anderson never had the feeling that Harvey expected or wanted his research to produce something. "He had no preconceived notions about locating a physical correlation to genius," he said. "His role was curator." Nonetheless, he included Harvey as co-author and sent him a draft of the three-page article. "I had this sense Harvey would feel good to know that the brain was worked on and that the findings would be published in a journal that you could see and read if you looked it up in a library," Anderson said. "I definitely consider him a collaborator . . . if he hadn't kept it, it might not have been kept at all."

Under the weighty title "Alterations in cortical thickness and neuronal density in the frontal cortex of Albert Einstein," the Anderson paper offered an honest summation of the project: "Studying the brain of a genius can play a small and titillating role in the quest to identify these neurobiological features that affect intelligence." Harvey read it carefully and phoned Anderson back to find out more about the reference citing the work of this Canadian scientist, S. F. Witelson. Anderson told him that Witelson worked at McMaster University and that she was well known and respected in the field of behavioural neurosciences, having made several interesting discoveries. She also had what might be the largest collection of normal brains in North America—if not the world.

Neuroscience Letters eventually printed Anderson and Harvey's Einstein paper in its June 1996 issue. The splash the young researcher had hoped for turned out to be little more than a splatter. About fifty scientists telephoned for reprints, and the local news ran a short item on the research. Otherwise, it did not open up Anderson's professional opportunities as he had hoped. The most remarkable reaction was a chilling one that came in a letter postmarked Georgia. The writer, who scrawled in German, pretended to be Einstein himself: "It included a threatening, disturbing passage asking that he be left in peace."

THIRTEEN

METAMAN

☼

IN HAMILTON, ONTARIO, in the fall of 1995, the fax machine in the Psychology Department of the McMaster University Health Centre spat a single sheet into its receiving tray. The third-floor warren of rooms usually buzzes with activity, and probably no one noticed its electronic babbling, the uneven staccato of ten shaky, handwritten words droning out in black ink. Students might have been filing in to gab with instructors and turn in assignments. The assistants and secretaries who sit behind desks in the common reception area were likely answering telephones for the professors in adjacent offices, typing reports or perhaps even tacking the latest press coverage to the faculty bulletin board, where a veritable collage of newspaper articles advertise the popular appeal of the research undertaken in this particular pocket of academia: clippings from the *Globe and Mail*, the *Toronto Star*, the *New York Times*.

The latest onslaught of media attention had been generated once again by a dynamo of a professor in the corner office, easily the biggest producer of headlines in the department. Sandra Witelson's report that women have more brain cells than men in the language area of their brains had been published that May in the *Journal of Neuroscience*. As potential fodder in the battle of the sexes, it brought national networks running. But by age fifty-five, the chic and petite Professor Witelson had grown savvy with experience. To many interview

requests, she'd shake her corona of raven curls and slice the air emphatically with small fluttering hands and say no—absolutely not. For
those she accepted, the conditions of her co-operation were carefully specified in advance. She had, after all, once been ambushed on
live television. Invited as a guest panelist to discuss her research on
the inborn differences between men's and women's brains, Witelson
had found herself on the defensive, accused of setting back the women's
movement by suggesting females are better suited to certain tasks
and males to others. "They had me on with Gloria Steinem," she told
the women's magazine *Chatelaine*. "It was obvious Steinem hadn't
read my work. She didn't know what I was about, what my work was
about. I felt very, very set up—thoroughly used. This is a subtle,
subtle, subtle area." Her daughter, a television producer, then gave
her some sage advice, and after that anyone who wanted airtime with
Witelson first had to surmount her wall of assistants. Often, they
were the ones who returned her calls, vetted her e-mail and picked
up her faxes. So it's a safe bet that someone else scanned that single
sheet first, arriving as it did that day without warning or cover letter. Then they likely scurried with it straight to her office. It would
always amaze Witelson how it landed in her lap like that, out of the
blue, with no introduction or preamble, no details or context. Only
the cramped scrawl of a single question—"Do you want to study the
brain of Albert Einstein?"—and beneath it, a signature. Who on earth
was Thomas Harvey? she wondered.

✸

The fuss over his 1994 public statement about finding a home for the
brain reminded Harvey how tired he'd grown of slitting open
envelopes, of separating the pleas of souvenir-stalkers from serious
requests—and there seemed to be so few of those these days. But if
the picture painted in the *Wall Street Journal* article was accurate,

nobody else was interested in taking on the responsibility of preserving Einstein's brain in the way he had. He must have shuddered
at what he read. The Wistar Institute for medical research at the
University of Pennsylvania in Philadelphia had once had a brain collection of more than two hundred specimens, including some that
had belonged to important, famous people. But tissue had been lent
out over the years, tossed out in some cases and in others inadvertently destroyed. A lab technician handling the brain of esteemed
American poet Walt Whitman had accidentally dropped it on the
floor. The post-splatter damage had apparently been too extensive
to leave any option other than discarding it. The Wistar collection
had dwindled to a mere eighteen brains, and they weren't interested
in storing the one that had belonged to Einstein. An official with the
Smithsonian Institution in Washington had said, "That's not the kind
of stuff we collect." Even old Harry Zimmerman was indifferent. The
Journal reporter had called Zimmerman, assuming that the Albert
Einstein College of Medicine would be interested in housing the brain
of its namesake. But just as the ninety-three-year-old neuropathologist had likely told Nathan a dozen years before, that wasn't necessarily so: "It's a normal, male, human brain. After that, there's nothing
more you can say. . . . What Harvey's got is no longer of any value."

Years ago, Harvey recalled, someone from the Harvard Brain Tissue
Resource Center had telephoned him to gauge his interest in leaving
the organ there. The brain bank's credentials could not have been more
impressive. Based at McLean Hospital in Belmont, Massachusetts, it
was the largest federally funded brain-tissue repository in the United
States. McLean itself is affiliated with Harvard University—which,
ironically, once stood as the only American institution of higher learning that did not clamour for Einstein's presence after he first emigrated to the United States. Officials planning the university's 1936
tercentary celebration chose not to invite the superstar physicist,
fearing that Einstein hysteria would steal the show from all the other

scholars. Francine Benes, director of the bank, could not recall ever speaking with Harvey, or wanting Einstein's brain, but she did say that McLean's would not have accepted the organ because of its limited use for research. "Apart from the cachet of it being Einstein's brain," Benes said, "there's not very much you can do with it." Harvey said it was he who chose not to send it to McLean's, for the usual reason that he "didn't know those people over there very well." But sometime around 1994, Harvey said, he actually wrote to the Hebrew University in Jerusalem, which housed the Einstein archives, and asked if they wanted the great physicist's brain. According to Harvey, they said they didn't. "Why, I was quite surprised to hear that," Harvey said. "I never understood why they would turn it down, why they wouldn't want it."

That same year, scientists who had studied Lenin's brain, working under the veil of Soviet secrecy for seven decades, publicized their less than spectacular findings. Oleg Adrianov, director of the Moscow Brain Institute, announced that the organ's anatomical structure contained "nothing sensational." Despite Oskar Vogt's breathless announcement that he had discovered a biological explanation for the Bolshevik leader's "multiplicity of ideas," a finding which had intrigued the young Thomas Harvey, Adrianov was now admitting that any attempts to link such observations to ability were nothing more than speculation, "a rude simplification." Lenin's brain might boast large pyramidal neurons or a bigger frontal lobe, the director explained, but no one knew if this bore any relation to Lenin's achievements. Still, Harvey was happy to learn that the Moscow Brain Institute nonetheless believed in the merit of storing Lenin's brain, slivered as it was by then into 34,000-glass-mounted wafers. The fallen Communist regime might have been preparing to pull Lenin's pickled corpse from its glass sarcophagus in Red Square, but Adrianov insisted that his brain slices should still be preserved. "They can be kept forever," he told the press at the time. "There is still much work to do. They can be examined in the future when science discovers new methods."

Doctors of the mid 1990s seemed more fascinated by the criminal than the genius brain. When the notorious London gangster Ronnie Kray died of a heart attack in prison in 1995, government pathologists removed his brain for study, without permission from the family. The scientists wondered if they would find evidence of psychosis in the brain of the sixty-two-year-old who, during the 1960s, had terrorized the East End's mean streets with his twin brother Reggie, building a bloody empire of gambling dens, extortion and murder. Nearly a year after his death, a British tabloid broke the news of "the great brain robbery," and Kray's stunned widow, best-selling crime writer Kate Kray, author of *Natural Born Killers*, said: "After twenty-seven years locked up I thought death had finally freed Ronnie. But now I'm outraged to learn there's a part of him still imprisoned in a jar with a little paper label. It's disgusting. . . . You'd think someone could have consulted with me or said, 'Listen, Kate, okay if we take Ronnie's brain for a bit?'" The brain was returned to the Kray family for burial in its own casket eleven months after his funeral.

The same year Ronnie Kray lost the contents of his head, the preserved brain of American serial killer Jeffrey Dahmer ignited an emotional court battle between his parents in Wisconsin. Serving a 936-year sentence after admitting to having killed, raped and mutilated seventeen men and boys in his Milwaukee apartment, Dahmer was bludgeoned to death with a metal pipe by another inmate. His bashed body turned up in the morgue of the University of Wisconsin Medical School, and the staff could not contain their curiosity. Just as doctors had filed into a Princeton autopsy suite in 1955, thirteen medical professionals crammed in to get a closer look at the man who sometimes froze and cannibalized his victims. "Such was the fear of

this man that chains were on the feet, even post-mortem," pathologist Robert Huntington noted. "With all this crew," he said of the
spectators, "it is interesting that we do still have room for the body."

When the autopsy was finished, Dahmer's brain was not returned
to its cranial home. Dahmer's mother, Joyce Flint, had asked doctors
to preserve her son's brain, hoping researchers would find within it
a tumour or a scar, some clinical explanation for his despicable deeds.
But Dahmer's father, Lionel, challenged the move in court, arguing
that any study of his son's brain would only prolong the anguish, "open
up an entirely new chapter and create a whole new life of its own."
He told the judge that his son had stated clearly in his 1993 will that
he wanted to be cremated. The judge sided with Dahmer's father in
December 1995 and ordered the brain cremated.

Forty years after the Einstein autopsy, as controversy persisted
over organs taken without permission, a group of rabbis wrote that
keeping the late physicist's brain after all this time seemed sacrile-
gious. They wanted it to have a proper Jewish burial. Yet even though
Harvey himself had become concerned about the brain's fate, he still
felt their request was premature. "The rabbis thought that his brain
ought to be handled by rabbis in the final disposition of it. I thought
this would be okay when all the research was finished, which it
wasn't, I didn't think," he said. "But by then a lot of it would be dis-
persed and there wouldn't be much of it left to bury." Even then, in
some untilled field of conscience, Harvey knew this was mostly wish-
ful thinking. What research? He had lost the energy long ago to lobby
scientists to study it, and somewhere along the way he'd lost the inclination as well. "I could be lazy," he admitted. "I was getting more and
more requests for interviews and for tissue and I thought I'd better
turn this over to some younger people." But one day melted into the
next, while Harvey, as he always had, simply waited for the right
opportunity to present itself.

✳

Gregory Stock could hardly get over it. "Amazing," he kept thinking, "it really was true." For years, Stock had half suspected that Rob Harvey was pulling his leg when he used to tell those stories about growing up with Einstein's brain in a jar on the mantel. It cut him up remembering the antics Robert had described when they were in Baltimore together back in the 1960s, Gregory studying biophysics at Johns Hopkins and Robert at the Maryland Institute of Art. Robert had said that he and the other kids used to take a running leap at that jar on the ledge and give it a whack just to watch the famous brain bob about. Who knew if it was for real? What's more, it had been such a hilarious story, who cared? But there it was, in the 1994 *Wall Street Journal*, a story about Robert's father—at least the name was the same—and the amazing brain. So, it really was true. Stock had to laugh. Imagine a brain like that, an unequalled cultural icon, just lying around, the embodiment of intellect in the modern era, serving no useful purpose. Preserved for what—posterity? From what he'd read, scientists didn't seem that interested in it. But Gregory Stock, a strapping forty-four-year-old with a trim, salted beard and hazel eyes, by any measure a Renaissance man with runaway successes in wild and varied endeavours, was definitely interested. This was unbelievable. Did no one understand the value of this? he thought. To the *Wall Street Journal* it was just a quirky story. But there had to be something worthwhile you could do with it, some way to capitalize on the preservation of this genius organ, he mused. Mental aerobics rarely let him down. He'd already earned a doctorate in biophysics and a Harvard MBA, developed electronic banking software and authored a series of phenomenally best-selling books. If Harvey was a cup of warm milk, Stock was a double espresso. If anyone could pull together a plan to make the most of Einstein's pickled brain, the Harvard Business School's 1987 Freund-Porter Entrepreneur of the Year was a likely candidate. "It reminded me of Jesus Christ and the pieces of the cross," he said. "Fake chips of them had been sold and resold through the ages for fortunes, and here was the real thing: the brain of Albert Einstein. Nothing could equal it. And if science was the modern surrogate for religion, this was a holy relic."

Stock had read the articles that depicted these celebrity brain specimens "as though they had no significance," he said. "Somebody had dropped Walt Whitman's brain on the floor!" He had grander hopes for Einstein's brain. "I thought this simply had too much power to gather dust somewhere. Just to touch it would be a profound experience for too many people." As an object of scientific study, its worth seemed negligible. But the public interest it attracted made it valuable, he thought. This brain, if handled properly, had the power to generate funds, money for a greater social good, for medical research perhaps. "Rather than denying that it has the appeal of a relic," he thought, "why not accept that and use that for a bigger purpose. In the process, it would be fun to force people to deal with their feelings about the human body, mortality and the legacy we leave."

Lately, Stock had been following the story of the Visible Human Project in the news, a remarkable government-funded undertaking to compile detailed three-dimensional images of the anatomy of a human male and female. Generations of anatomists had struggled to get a good look inside parts of the human body, but how could they let in the necessary light, or find space to manoeuvre, without disturbing everything else around it? For centuries they had to be satisfied with dissecting particular parts from different angles and different depths. But not any more. Computer technology made it possible to take a body, slice it into thousands of cross-sections, photograph it from all angles and digitize it so that it could be manipulated on a screen.

The trickiest task proved to be finding the right bodies. Most cadavers were too old or diseased and had to be excluded as candidates. Then, in 1993, scientists at the University of Colorado, one of three schools to win a share of the federal contract, picked a male finalist: a thirty-nine-year-old on death row, a fitting choice considering that the medical field has historically relied on the cadavers of criminals for its raw materials. After failed appeals to the Texas Supreme Court, Joseph Paul Jernigan had agreed to donate his body to science after his execution for murdering a seventy-five-year-old man during a robbery in 1981. Researchers scanned Jernigan's cadaver using a variety of techniques, froze it in gelatin, and then literally milled it horizontally, one dime-thin slice at a time, photographing each newly exposed surface with a digital camera so that the pictures could eventually be stacked into a computer model. Why not do something like that with Einstein's brain, Stock wondered: cut full, thin, serial sections of the organ, compile computer reconstructions of the fine details of the structure, then mount the actual slices in a tasteful plaque or sculpture and sell them, or perhaps they could be individually numbered and leased with the stipulation that they could be recalled for scientific exhibition. You could probably earn enough to fund a large-scale, non-profit medical research project, he thought, something with broad appeal.

Stock raised the prospect with some of his art dealer friends. They estimated that a single slice of Einstein's brain could fetch $25,000 from museums or individual collectors—a person would have to sell only four hundred slices to raise $10 million! He thought he could probably sell several thousand pieces and earn $50 million or more on this limited-edition specimen. Not that he or anyone else would line their pockets with the proceeds of selling Einstein's brain, he said: "We'd be viewed as a bunch of charlatans if we did that. But if it was squeaky clean, people would not be able to pigeonhole it or discount it."

Before he got too carried away with the potential of his plan, Stock figured he'd better talk to Robert Harvey. They'd kept in touch over the years. Robert had moved back to Princeton to look after his mother, Elouise, and that summer, Stock, who was usually based in

Los Angeles at the University of California, happened to be a visiting fellow at Princeton's Woodrow Wilson School of Public and International Affairs. Of course, when it came to Einstein's brain, he and Robert had never before shared anything more than a belly laugh.

Robert Harvey laughed again when they met, this time at the sheer chutzpah of his old friend. Stock wanted to know if his father had made any plans for Einstein's brain, and Robert said he didn't really know, though he rather doubted it. Robert was, in a weird coincidence, hired in 1978 to help craft the clay armature of the Albert Einstein statue outside the National Academy of Sciences in Washington, D.C. (the fact that he had grown up with Einstein's brain in his house had been a useful credential). He sometimes worried that his father would pass away and the brain would just be left behind in a box somewhere. "He never really talked about it much, and we just didn't know what would happen to it. I guess we thought it might just be forgotten," Robert said. He hadn't spent much time with his father after Harvey had left Princeton. But like the rest of his siblings, Robert, who had inherited his father's soft speech and gentle manner, felt his dad had been given a rough ride over the brain, that he had been wrongly depicted as a thief. Robert had never heard his father rage about any of that. Then again, he'd never heard him rage about anything. "If he ever had harsh feelings about anybody he never said anything out loud. I don't remember him ever getting angry. He'd grimace and go quiet, but that was all," he said. "Even some people I met years later who said they knew my father told me that he always kept his counsel, the school superintendent and the owner of the hardware store and a psychologist he knew . . . they said they never really knew him and I don't think I knew him." When Robert did see his father he would tease him good-naturedly once in a while. If Harvey announced intentions to do this or that, Robert would say, "Gee, Dad, I hope it doesn't take as long as your study on Einstein's brain." But then he'd throw his arm around his father's hunching shoulders and give him a playful squeeze.

Robert agreed to put Stock in touch with his father. He said his dad was "surprisingly open to new things." Besides, Robert thought his friend's plan sounded intriguing. After all, Stock's ideas had made him rich and famous—and he'd sown the seeds of his celebrity while merely gabbing with a woman in an Oregon café in 1984. Stock had started toying with the provocative questions to spark conversation. They discussed raising prickly ethical issues and odd dilemmas about love, values, money, priorities and sex: "Would you rather be happy, yet slow-witted and unimaginative, or unhappy yet bright and creative?"; "Would $100,000 be enough to put a loyal, healthy pet to sleep?"; "Would you personally murder a single innocent child if it would permanently end world hunger?" Stock sensed he'd hit on a marketable method to inspire meaningful, or at least interesting, conversation. He and a friend ploughed $10,000 into publishing a hundred-question pamphlet. Eventually, he hooked up with a New York publisher, and in 1987 his quirky, provocative *Book of Questions*, which contained not a single answer, shot to the top of the *New York Times* best-seller list and stayed there for twenty-six weeks. While foreign publishers translated it into seventeen languages, and more than two million copies were sold, Stock, whose earlier publications had been academic papers on amphibian limb regeneration, bacterial motility and three-dimensional computer reconstruction, churned out three more books of questions, one for kids, one on love and sex and a third on business, politics and ethics. People said he had plugged into the moral ennui of the excess decade.

In 1992, Stock ran for a seat in the U.S. Congress. In 1993, he published his fifth book, *Metaman: The Merging of Humans and Machines into a Global Superorganism*. The critics called it visionary, an unsettling but optimistic thesis that humans are shaping their own evolution with technology. Glued together by trade and telecommunications, the people of the world were joining together to form a single immense organism, Stock wrote, criss-crossing the planet like cells in a larger organism. It was his work on Metaman that led him to consider the scientific potential of the revolution in genetics and the prospect of manipulating human biology in profound ways. It was feasible, he thought, that in the near future scientists would be able to insert genes in human chromosomes to retard the aging process. Ultimately, Stock decided that using the sales of Einstein's brain to fund research into aging would be the ideal project. Such research, he thought, would have mass appeal. Who wanted to grow old? Einstein himself had lamented his physical powers fading with age. But he had also written at the age of fifty-seven that "I live in that solitude which is painful in youth, but delicious in the years of maturity."

☼

Robert knew that his dad heard from quite a few strangers about the brain and thought it best to pass Stock's number on so that Harvey could call him. Eventually, Stock sent a letter to Lawrence, Kansas, and soon after the two men spoke on the telephone. Harvey said he was interested in hearing more about the proposal. On August 4, 1994, Stock laid out his proposition—with conservative financial estimates—in a clear, four-page summary, complete with two sub-heads: "The Goal" and "Raising The Funds."

Dear Dr. Harvey,

It was nice to finally be able to talk with you about the concept I'd alluded to. . . . I'd like to use Einstein's brain to raise enough funds ($10 million or more) to found a unique and innovative foundation that could help address the important medical-research problem of aging. The money would be raised by offering 1,000 sculptures (a series of identical pieces, each an homage to Einstein centered around a stained and mounted thin section of tissue from his brain) to contributors of at least $10,000. . . . [T]he project would be a sincere tribute to Albert Einstein, the end result would further basic medical research and benefit the public, the use of

the tissue would not be commercial, and the conduct of the publicity and fund-raising would be above reproach. Something like
this would be subject to enormous scrutiny, and the outcry would
be deafening were the design and execution of the idea not genuinely worthy.

Stock felt he should stress that no one involved in running the
project would personally profit from selling Einstein's brain. "I would
not have taken a penny," he said. "The project itself was what interested me." But raising the spectre of hucksterism could actually work
in the project's favour, Stock thought. He wrote:

Einstein is an icon that has come to represent the jump in scientific
understanding that brought humanity's transition to the modern
world. You've seen, Dr. Harvey, how much interest was generated simply by the idea that Einstein's brain was in your possession. Well imagine the stir that would result from what would be
seen by many as the sale of his brain. Some, of course, would be
offended by this, but the motivation, goals, and execution would
be pure enough so that such critics would be hard pressed to make
a case. And remember, not only is embedded body tissue routinely sold through scientific catalogs for use in anatomy classes,
but the traffic in holy relics (bones of the Saints and such) has a
long tradition. I cannot deny that there's a certain macabre aspect
to such relics, but there can be no doubt that they inspire people
by giving them the feeling of touching greatness (or holiness),
and I suspect that these mounted sections of Albert Einstein's
brain would affect their owners in the same way. For many years
I kept a bust of Beethoven above my desk to remind me to strive
for greatness, and a slide of Einstein would have inspired me even
more. . . .
    I already have ideas about how to incorporate tissue sections
into an homage to Einstein. Basically, the stained tissue would be

central but coupled with a motif evoking Einstein's influential [1905] papers on relativity and the photoelectric effect. The piece would be simple, uncluttered and powerful. Finally, I've discussed with several art dealers the sale of a limited edition of such a piece and am confident that the kind of sale I've suggested is feasible. There are good ways both to stimulate demand and gauge the market strength so that pricing can be optimized. . . .

As to how we might proceed—if we do—I'd need your formal permission to go forward . . . but I'd very much appreciate it if you'd agree to be available as an ongoing consultant about the tissue samples and history of the material. And it seems to me that compensation of around $50,000, payable from the money raised would be appropriate for that service. As to myself, I'd eventually be reimbursed for any outlays in setting this up, but my goal is not to earn income off this but to give life to the research foundation I described. . . .

In closing, I want to emphasize how extraordinary an opportunity this is—one that you made possible by preserving the tissue from Einstein's autopsy for four long decades. Now the direct scientific usefulness of the tissue itself is likely rather marginal, but its symbolic value is not and can be harvested. This is an audacious, outlandish idea, I know. But it is one that can succeed, so I hope you will consider it carefully. The risks are small, the potential rewards large, and I suspect that Einstein himself would not be offended by what I've proposed. After all, it's a far more fitting close to the unlikely tale of Einstein's brain than some dusty museum closet.

Sincerely,
Gregory Stock

Stock flew to Kansas in the fall of 1994. For a man of eighty-two, Harvey appeared unusually energetic. Stock was impressed. They shook hands and chatted in the crowded space of aparment 13, the wide-eyed visionary and the passive pathologist. They discussed Stock's idea, and Harvey wanted to know whether researchers would still be able to investigate Einstein's brain if they sliced it up and sold the pieces as souvenirs. Stock assured him that research could continue on a computer-reconstructed model of Einstein's brain, in the style of the Visible Human Project. According to Stock, Harvey seemed quite willing to go ahead with the plan if those were the terms. "I sensed that he would have been glad to be involved in that kind of a scientific project," Stock said. "He got excited about it." Then the moment arrived, and Harvey led Stock to his bedroom closet. Like waves of people before him who would never forget the instant they set eyes on Einstein himself, Stock felt slightly overwhelmed at the sight of the soft-tissue remnants of the century's great genius: "It was surreal." Harvey stooped over into his closet "to get it from behind some socks or something." Then, just as he had done for the fanatical Japanese math professor, Harvey picked up the glass containers and carried them to the table. "It was so weird," Stock said. "He had this thing in a bottle in a closet and he brought it out and stuck it on a cheese board." As Harvey dipped his fingers into the hazy liquid, hoisting thumb-sized pieces to the surface and out into the open air, Stock realized that he had horribly misjudged the state of the organ. When Robert Harvey had told stories about it, Stock had envisioned a whole brain sloshing about in its bottle on the mantel or, at worst, a brain in a few pieces. But not this, "not a couple of hundred fragments."

Suddenly, the prospect of cutting large serial sections of saleable pieces presented a much grander challenge. As they sat at the kitchen table that afternoon, with the bits of Einstein's embedded brain on the cheese board—"We were fiddling with these little yellowed plastic chunks, each containing a piece of Einstein. . . . I was sorely tempted to take one"—he asked Harvey if all the pieces were there. Harvey said he had "given a few away." Stock told him that he would have to investigate the possibility of reconstructing the organ as a whole; producing full serial sections of the organ would not be possible otherwise. "I wanted," Stock said, "to put it back together again."

Stock eventually spoke with two scientists willing to devote their time and resources to reconstructing Einstein's brain, Charles Ide, a biologist from Tulane University in New Orleans, and Wally Welker, a neuroanatomist at the University of Wisconsin. The experts estimated that to melt the celloidin from the tissue and painstakingly glue the two hundred-odd remaining pieces back together, with an adhesive strong enough to withstand cross-section microtome slicing, would cost at least $100,000. They never even considered trying to include the slides, since Harvey had lost track of half the boxes. But the important thing was that reconstruction was still possible. Stock liked the challenge. He thought he could create a little book and call it something like *Reassembling Einstein* to document the effort.

Stock called Harvey before Christmas 1994 and told him it would probably be a good idea if he visited the scientist in New Orleans who would prepare the tissue, which would later be reassembled by Welker, a master of the art in Wisconsin. Harvey said he could travel early in the new year, and Stock said he would make the arrangements and cover the costs of the trip. On January 3, 1995, he sent the itinerary. The plan was for Stock to fly to Kansas so that the two of them— with Einstein's brain as carry-on—could head down to New Orleans and leave the brain with the scientist at Tulane.

So this was it. This was how it was going to end for Thomas Harvey. He'd leave Einstein's brain in New Orleans, spice, sax and celebration city. Just drop it off and fly home. After a lifetime of alcohol refills and mail and media, of quiet moments admiring the slides and the golden encased nuggets of brain, the bond that had lasted longer than any of his marriages would be broken. His custodial services would no longer be required. Harvey had the cold feet of a nervous groom. He called Stock in California as soon as he received the letter.

A trip to New Orleans in January did not suit his schedule, he said. Well, when could he go? Stock asked.

Maybe he could make it down to Tulane in February, Harvey replied.

Stock needed to know a date so that he could buy him an airline ticket.

That wasn't necessary, Harvey said, he'd prefer to take a nice slow drive.

When Stock wrote to Harvey two days later, on January 5, he said he'd arranged for Harvey to visit the Louisiana lab by himself in mid February. "Hopefully," Stock wrote, "we'll be able to begin moving forward in March with the reconstruction of the tissue. I'm eager to get started, because the process of assembling the brain, sectioning and staining the material, collecting data . . . will likely take at least a year's effort."

Stock sensed Harvey's growing reluctance to part with the tissue, and toward the end of the month they spoke on the phone again. It was true, Harvey said, that he had concerns about the project and the appearance of propriety. He didn't want to be accused of exploiting Einstein's brain, to be subjected to the same sort of publicity that surrounded Henry Abrams and his possession of Einstein's eyes. The former Princeton ophthalmologist, Harvey said, had telephoned him some years after the autopsy and confessed that he had taken them. Abrams kept them locked in the safety deposit box of a bank on the east coast of New Jersey. He still gazed regularly at the inky centres of the two white globes, half nestled in their soft tissue pouches, cramped, one on top of the other, translucent nerve strands floating gracefully out behind them in a formaldehyde-filled jam jar. Perhaps it was in a flush of guilt that Abrams offered Harvey his prized possession. Harvey said Abrams hoped the eyes might be of some use in studies of the brain, that they might have some worth beyond his veneration of them. Harvey told him he didn't want them: "I didn't know he had taken them and I was unhappy to hear about that."

Harvey bristled whenever talk of the eyes popped up, loathing the idea that he might be lumped together with Abrams. When it had all come out in the newspapers the year before, in 1994, Abrams was sometimes quoted as saying that Harvey had actually been present when he'd scooped the eyes from Einstein's empty head. Harvey always maintained that he wasn't, that he was outside on the hospital's front steps, holding a press conference. Then rumours trickled through the press that the retired ophthalmologist might sell the eyes for $5 million. It was said that pop music star Michael Jackson, who had once bid for the bones of the so-called Elephant Man, was interested in buying. Abrams vehemently denied talk of a sale and condemned the tabloid coverage of his self-described act of adoration. "I don't know how that ever intruded into the story," he said.

Stock had no doubt that the publicity surrounding Abrams had "spooked Harvey a little bit." But the newspaper stories had been out for a year, months before the two of them discussed their plans for Einstein's brain. Why had Harvey not mentioned earlier that it bothered him? The scientist-cum-businessman tried to convince Harvey that he was a pro at handling the media; he told Harvey that he had been on more than a thousand television and radio shows, including *Larry King Live* and *Oprah*. It seemed to Stock that the real issue had little to do with public perceptions. The fact was, the closer they drew to the date on which Harvey would actually have to give the brain up, the less inclined the elderly pathologist was to participate. "Harvey probably realized he would not be in charge," Stock said, "and he began to feel some anxiety at the idea that he'd actually have to let it go." Stock worried that he might expend enormous effort to co-ordinate the project, recruit other scientists to join the team, and then what? "I would be dealing with this unpredictable old man," he said. "I could never make representations to anyone about it."

On January 28, 1995, Stock wrote Harvey a final letter. Sensitive though he was to his feelings about Abrams, his comments cast a

stark light on the alternative future of Einstein's brain if Harvey
reconsidered their agreement.

> Dear Thomas,
>   . . . I want to comment on the various reports about Dr. Abrams.
> I see no similarity between what we're doing and what he's doing
> . . . . Our situation is entirely different. . . . [Y]ou were autho-
> rized to perform the autopsy, and you removed the brain with
> the eventual permission of the family. . . . [Y]ou've been scrupu-
> lous about allowing only scientific research on the brain, and have
> been selective even about that . . . [T]his project has a scientific
> justification that will be corroborated by major researchers work-
> ing on computer reconstructions of the brain. Any reporter will
> be able to verify this. . . . [W]e are not receiving personal gain
> from this project. Yes, money is being raised, but only to support
> scientific research including that on the brain itself. Moreover, a
> non-profit foundation (with financial records in the public domain)
> will control the funds, so any double-dealing would be a crimi-
> nal offense!
>   Undoubtedly, you can find others who would be willing to take
> Einstein's brain and putter around. But that has been the case for
> 40 years, if I understand correctly. . . .
>   If we don't go forward, I imagine the brain will either end up
> tucked away in a dusty storeroom of some museum, or else some-
> one will get ahold of it after you die and it will end up being
> destroyed with no serious science ever being done. What a tragedy
> that would be. . . .
>   Thomas, you've faithfully watched over this brain for 40 years
> now—protecting it from specimen hounds and memento seekers,
> keeping it largely intact. Now technology has reached the point
> where some interesting science might actually be possible. I'm con-
> vinced, however, that it will require the type of approach we agreed
> to pursue last summer. . . . We can create a legacy worthy of

Einstein, and one that will bring you credit too . . . please think about the likely alternative—that nothing very interesting will happen and that Einstein's brain will remain unstudied, at least for the foreseeable future. . . . I've now brought together all the pieces necessary for success. Please don't let your second thoughts undo things.

Of course, you don't *have* to do this—we both know full well that your say is the final one—but please allow yourself to imagine the tremendous possibilities of what we've discussed. And remember, you eventually will have to give up control. So, why not be a part of something exciting like this.

Regards,
Gregory Stock

February came and went, but Harvey never left Lawrence. He said something had come up. Stock stopped calling and writing. For Harvey, Stock realized, "it was just too psychologically difficult to give up." There he was, "living like a student in a tiny apartment," obviously in need of money, but he was unwilling to play the role of paid consultant because it meant the tissue would no longer belong to him. "It was so central to his life. This strange affair with Einstein's brain thing turned out to be the most important thing he'd done in his life really and people from all over, from the *Wall Street Journal* and from Japan and all over came to talk to him about it," Stock said. "It was his connection to something larger than himself. It had become the meaning in his life."

When Stock told Robert what had happened, they thought perhaps someday, if Harvey left Einstein's brain to his children, they might pursue the plan then.

Harvey never denied it. Stock was right, he'd say years later: "I wasn't ready to give it up, I guess, at least not for his project." Although he had never shared much more with the living Einstein

than a trivial conversation, Harvey proved no different from Otto
Nathan or Helen Dukas—bound, driven to maintain control over
Einstein's earthly legacy. Harvey backed out of Stock's ambitious
scheme, he said, because "it was going to be a little too commercial.
I knew that Otto Nathan and the Einstein family, his descendants,
were opposed to commercializing Einstein." Yet Harvey had no
idea who among Einstein's relatives was still alive or what they
might have thought of Stock's plan. He was not aware that they had
no rights whatsoever in the very deliberate commercialization of
their famed family member's estate. Three years before he died,
Otto Nathan had initiated a deal for the Roger Richman Agency in
Beverly Hills to license Einstein's image and funnel the fees to the
Hebrew University.

Roger Richman's father, Paul Richman, a rabbi and director of
the Anti-Defamation League of B'nai Brith, had worked with Einstein
in 1938 trying to get Jews out of Germany and finding them jobs in
America. After Einstein died, young Roger Richman used to travel
with his father to Nathan's New York apartment. Richman was around
twelve at the time, but the memory of the elderly Nathan in his airy
study surrounded by Einstein's books and papers stuck with him.
After Roger Richman became a lawyer and made a foray into films,
he set up an agency to represent the families and estates of dead
celebrities, to prevent the unauthorized use of their names, photographs and signatures. He inspected flea markets and five-and-dimes,
nixing John Wayne toilet paper ("It's rough, it's tough—it doesn't take
crap off anyone") and the supposed bottled sweat of Elvis Presley
("Perspiration for inspiration"). Eventually, Richman launched a
crusade to halt the exploitation of "dead legends," as he calls them,
and claimed a victory when he authored the California Celebrity
Rights Act.

In 1984, in the midst of preparing his case for the state legisla-
ture, Richman began clipping the numerous ads that relied on the

unauthorized use of Einstein's image. It was well known that the Nobel physicist had abhorred the notion that his persona might be used as a promotional gimmick, but after his death the sentiment didn't count for much. Einstein's metaphorical mane appeared in promotions for everything from hair salons to General Electric. "I tore them out and sent them to Nathan," Richman said. The agent eventually told Nathan that Einstein was in such hot demand that the estate ought to have some control over the appropriate use of his image and receive some benefit for allowing it. For Nathan, as always, exercising some control was better than no control. He put Richman in touch with the Hebrew University. To this day, Einstein ranks as Richman's hottest property. The late physicist has generated more in licensing fees than John Wayne or W. C. Fields. While Richman has halted plans for Einstein tobacco, Einstein vodka, Einstein brain supplements and slogans that feature "e" equaling anything but "mc^2" ("We never allow a company to replace his actual chalk marks with their own logo," he said), the image of the dishevelled physicist has been approved to sell a Japanese video game, a respiratory antibiotic, the services of a South American telephone company, Apple Computers (the "Think Different" campaign), Nikon cameras, Coca-Cola and Pepsi. There are authorized Einstein ties, mugs, pencil holders and squishy Einstein dolls.

When Harvey learned all this years later, he commented that an image is "not the same as commercializing his brain. . . . His brain was only supposed to be used for science." But there could be no more procrastinating. The echoes of Stock's last letter, which Harvey had saved like all the others, must have resonated in his mind: *the brain will end up in a dusty storeroom . . . destroyed with no serious science ever being done . . . Einstein's brain will remain unstudied. . . .* No more could he treat the future as though it were a distant sunset he could meander toward in his own sweet time. The future had arrived. It was 1995, the fortieth anniversary of Einstein's death.

If Britt Anderson thought Harvey sounded glad to hear from him when he telephoned from Birmingham that spring, it's because he was. Harvey took the call as a hopeful sign. He'd mailed a few slides away to researchers in Germany, Japan and South Africa, but he'd never heard anything back. So it had been ten years since he'd last enjoyed the prospect of a scientist wanting to conduct a serious study on Einstein's brain. The notoriously absent-minded Harvey never even waited for Anderson to send the Federal Express package before he bagged the pieces and dropped them in the mail.

Then, that summer, while Harry Zimmerman lay dying in a Bronx hospital, Harvey heard about the work of Sandra Witelson and the big collection of brains she had up in Canada. The old pathologist decided to do something he hadn't done since he was a young man with a straight back, thick hair and ambition: *he* would solicit a scientist to study Einstein's brain. Maybe he thought he could disprove Gregory Stock's predictions. "I guess I really hadn't done that [pursue a scientist] since the beginning," he said. "I looked up one of her papers in the literature . . . I don't remember which one, I think it included her fax number at the end." Harvey took the number to work with him and, in a spare moment, he penned his succinct invitation on a single piece of paper and fed it into the office fax machine.

FOURTEEN

CANADIAN CARTOGRAPHER

✧

HARVEY DROVE AN old Dodge, Ontario-bound, on a saddle of wooden beads in January 1996, his shrunken frame bent over the steering wheel. Smoked fish and a loaf of bread lay scattered across the passenger seat beside him; the pieces of Einstein's brain sloshed about in their two glass jars in the trunk. He was nervous. "I had never brought the brain across a border before," he said. How would the customs officials react?

The first time he'd visited Sandra Witelson, in the fall of 1995, he'd left the brain at home. He'd said he wanted to make her acquaintance before he made a delivery. Home was still in Kansas then, and he had driven two days straight, north through Iowa and Michigan, over Detroit's Ambassador Bridge into Windsor, then east to Hamilton, forgoing the expense of roadside restaurants and motels for packed lunches and naps in the car.

When he'd returned to Lawrence, Maurellis had told him that they were going to have house guests. Anti-apartheid sanctions had made life too difficult for his family in South Africa. Maurellis's mother and grandmother were coming to America to live with them for a while in their two-bedroom apartment. Harvey felt it was time to leave. He packed his books and journals, his tool box and the brain, threw his arms around the hulking Maurellis, and promised to keep in touch. As much as he loved the Midwest, Harvey always figured he would end up back east. His children were there. Frances lived in Maryland at the time, Elizabeth in Washington, D.C. Thomas Jr. and his family lived in North Carolina. Arthur worked for the New Jersey Health Department and Robert and his wife lived with Elouise in their old family house on Jefferson Road. He would move back to Princeton.

Elouise Harvey had never remarried. She dated a prominent Princeton psychology professor for some time but nothing came of the relationship. Robert half suspected that his mother never stopped loving his father. In the basement, she still kept her old scrapbook of articles and black-and-white photos from happier days: with the boys, smiling and squinting into Harvey's Exakta on Jersey's Normandy Beach; Harvey posing with his lab assistants for the local paper; her wearing pearls and holding the Bible for his swearing-in on the State Medical Examiner's lab committee. And there too, frayed and yellowed, were the clippings from Harvey's first days with the brain. You had to hold the scrapbook gingerly to look at them; the pages had separated from the spine, and the whole thing fell apart if it was tilted upright. Robert remembered that his father showed up at the house soon after he moved back to New Jersey and asked his eighty-seven-year-old mother to consider a reunion. Elouise still had a stubborn streak, Robert said. "She asked him to leave."

Rejection proved to be only a minor setback. Harvey rarely had trouble attracting the affections of a good woman, even one twenty years his junior. The handsome Cleora Wheatley, a smart, no-nonsense sixty-seven-year-old divorcee with blue eyes and a white cat, took Harvey in to share her cozy bungalow in the tree-covered hills of Titusville southwest of Princeton. Wheatley had been a nurse at Princeton Hospital in the 1950s, and she was a partner in the ill-fated nursing-home venture Harvey and the Fishmans undertook in the 1960s. So she understood Harvey's history, appreciated his fine manners and Quaker ways and protected him like a Helen Dukas from the prying press. God help any reporter who turned up on

Pleasant Valley Road unexpected. "I know all the tricks," she'd holler. "People come here trying to catch him out, think they're going to find him in the raw . . ." She suspected that the media wanted to unmask "her Thomas," hoping to find something sinister beneath the elderly doctor's gentlemanly veneer.

Everything about their living arrangements suited Harvey. If Lawrence had been penance, this, surely, was absolution. In the summer, the sweet smell of tilled soil wafted up the hillside from farmers' fields flanking the roads below. In the clearing out back, Harvey could chop wood for the fireplace or the workshop in the basement, where he finally had enough room to store his boxes of books and brain ephemera. Cleora cooked his lunch, and sometimes old friends like the Fishmans came to visit. On Sundays, after his Quaker meetings, Harvey could sit out on the deck, pat the neighbour's golden retriever, which regularly bounded on to their property, and read the newspaper: retired at last. For a history buff like Harvey, Titusville seemed the perfect town as well, home to nineteenth-century architecture and Washington Crossing State Park, where Americana enthusiasts dress up in Revolutionary war costumes every Christmas and row across the Delaware. From Titusville, he could drive to Princeton in under half an hour, catch a quick train to the Quaker Information Center in Philadelphia, and, if he got an early start, he could reach the Canadian border at Niagara Falls in a day.

Well past midnight, Harvey rolled up to the Rainbow Bridge. He decided to pull over in case his cargo created a fuss: "I thought I'd better stop at customs and tell them I had this brain."

You can imagine the uniforms on the late shift leaning into his window, braced for drunken bar-hoppers from over the river, drug-smugglers or day-trippers in discount Calvins who tossed their old jeans at the factory outlet and stuffed their trunks with cheap booze. "Where are you coming from?" they ask officiously, sizing up a driver to judge whether he matches his story. "How long are you staying? . . .

Anything to declare?" Did they take one look at this earnest old doctor
who claimed to have in back the genius brain that changed our perceptions of space and time, and dismiss his story as a harmless crock?
Were they just too tired to be amused and happy to wave him through?

"They didn't ask to see it. They pretty much took my word for it.
They didn't seem upset or curious about it or anything," Harvey said
later, laughing incredulously at his adventure. "I was surprised to get
it into Canada so easily."

So north he went, relieved and happy, the eighty-four-year-old
man with Einstein's brain, cruising past the tawdry wattage of Niagara
at night, past street vendors' fudge and candy-apple carts, Louis
Tussaud's Waxworks museum and Ripley's Believe It or Not.

Harvey looked forward to seeing the Witelsons again. On his first
visit, they had taken him out to dinner and shown him some of the
local sights. "Henry Witelson was a very fine ophthalmologist and a
very kind man," Harvey said. He'd examined Harvey's eyes at his
office the last time he was there. "He found some problems, but nothing I didn't know . . . you know, the partial blindness that comes with
age." Sandra Witelson impressed him too. With her dark features and
catwalk cheekbones, she was a looker, Harvey thought, and a heck
of a smart scientist to boot—"a firecracker." Who could be better
equipped to spot differences in his remarkable specimen? Witelson
had the brains of more than a hundred people. She had plaster casts
made of each of them and stored the tissue in a collection that earned
the envy of scientists across the continent and beyond.

✵

By the 1980s, the supply of human brains fell seriously short of
demand. Research in genetics had raced ahead, sparking interest in
the molecular biology of brain disorders. Although scientists had been
describing the anatomy of Alzheimer's and other forms of dementia

since the turn of the century, they were now capable of dusting for the chemical fingerprints a culprit gene leaves at the scene of a diseased brain. In the gruelling, decade-long quest for the gene behind Huntington's disease, for example, researchers scoured patients' post-mortem brain tissue for a telltale protein. Too much or too little of a protein is a clue to finding the faulty gene that produced it. But when a brain is stale, the chemical evidence can be tainted. So the tissue had to be well preserved, preferably frozen, fresh from a corpse. Through hospitals and patient-support groups for conditions like schizophrenia, Parkinson's disease or amyotrophic lateral sclerosis (Lou Gehrig's disease), scientists at least knew where to go to ask for diseased brains. Relatives often saw the upside of donating if it would help researchers find a cure for a condition they might inherit. Next of kin sometimes delivered a brain to the tissue bank themselves. But a stash of diseased brains meant little if scientists had no healthy brains to compare them to—and they didn't. Normal post-mortem brains had become as scarce as the spotted owl. The boom in brain research had run up against a bust in autopsies, and fewer post-mortems meant fewer chances to harvest the resource inside cadavers' heads. Gone were the heydays of a Thomas Harvey, when even a smalltown pathologist could dissect nearly half the patients who died at his hospital and remove some sixty brains a year; gone were the flush semesters of Marian Diamond and the diener. Hospital autopsy rates in the United States plunged from a peak of 41 percent in 1964, to 15 percent in 1983, to less than 7 percent by 1995. Surveys showed a similar decline in Europe and in Canada, where, by 1996, only one in ten hospital deaths was followed by an autopsy.

With CT scans and magnetic resonance imaging machines, doctors felt they no longer needed to take an intimate peek at a dead patient's innards. They could glance at a screen and diagnose disease in the living. At the same time, hospitals under pressure to slash spending discouraged routine autopsies as an expensive frill: patients don't pay for them; most health insurers don't cover them. What's more, there's always the risk that an autopsy will turn up something a doctor missed or misdiagnosed and leave the institution vulnerable to a malpractice suit. Medical schools, meanwhile, spent less time teaching post-mortem techniques. Rudolph Virchow and Karl von Rokitansky could easily have been mistaken for concert pianists.

To make matters worse, brains tend to rank low on the list of organs a family will donate for medical research. A heart is a pump; a kidney is a filter; a lung is a ventilator—but what is a brain? If it's the mind, then does it not contain the essence of us, is it not the address of the soul? Relatives prefer to make a gift of Aunt Betty's liver than Aunt Betty's brain. "It's the place where Aunt Betty lived," as a tissue bank official once put it. "The brain *is* Aunt Betty."

So severe did the brain shortage become that researchers studying tissue for explanations of psychological problems began to have papers questioned because they could not collect enough brains to make their cases statistically significant. Scientists, like their post-body-snatching compatriots at the turn of the century, began talking about using their own brains. In a 1989 letter to the *New England Journal of Medicine*, two doctors from the brain tissue bank at the Harvard-affiliated McLean Hospital appealed to fellow physicians to support their efforts "by considering seriously the possibility of donating their brains for research after death." Anyone with their own steady supply definitely had bragging rights.

✲

When she received the mysterious fax with its remarkable offer, Sandra Witelson did not hesitate. Whoever Thomas Harvey was, he had Einstein's brain and an invitation too intriguing to resist. Maybe memories of her father popped into her head, flashes of her childhood home in Montreal. Her dad was a huge Einstein fan. He used to stock the study shelves with books on relativity. Although he was an accountant, he longed to be a physicist. People said her father even looked like Albert Einstein. How proud Hyman Freedman would have been to know that the eldest of his two children had been invited to study the great physicist's brain. Witelson scribbled a "Yes" on that same sheet of paper and faxed it back to the curious Thomas Harvey, out there somewhere, with the brain, in the electronic abyss.

Not until she went home that night to her husband in a posh Hamilton neighbourhood did second thoughts begin to nag at her. Perhaps she had been too hasty in agreeing, she told Henry. Imagine the crazy reactions she might elicit by studying Einstein's brain. Hadn't she already suffered her fair share of undesirable feedback from strangers, particularly in the months after she'd published research in the early 1990s suggesting that gay men think differently from straight men? But she couldn't deny that it sounded like a tremendous opportunity. For nearly thirty years, she had devoted her career to studying the relationship between behaviour and brain structure, and here was Einstein, the extreme example of an intelligent individual. Perhaps something in his brain could highlight a relationship between mental ability and biology.

So much of her relationship with Henry was rooted in a common passion for their work. They had been married thirty-five years, high school sweethearts from the Strathcona Academy in Montreal. Henry had been chief of eye medicine and surgery at the Hamilton Civic Hospitals since 1978, a gregarious doctor who built an enormous clinical practice and a reputation for having the kind of deft hands that attracted patients from far and wide. Whenever friends or reporters asked Witelson how she managed to nurture such an impressive career and raise a child at the same time, she credited Henry, a man she described as being ahead of his generation when it came to supporting his wife's professional pursuits. And that evening, with a mind-boggling offer on the table, Henry told her to forget the second thoughts: out of all her colleagues in the field, she had been chosen to analyze Einstein's brain. If she turned down the chance, he said, she would always wonder what she might have discovered. Besides, she had never been one to shy away from prickly research subjects.

✷

Sandra Witelson had a knack for hitting hot sociopolitical buttons. The results of her work were the kind that provoked strong reactions when reduced to sound bites: male brains are different from female brains; a part of the male brain shrinks with age; gay brains are different from straight brains. Where the Britt Andersons of the world ran up against the taboo of race and intelligence, Witelson tangoed with the tender issues of gender and sexual orientation. She always insisted that that had never been her initial intention. It had begun quite innocently— accidentally, in fact. But then she was exceptionally well equipped for the task, and her timing was impeccable.

While psychology students at most North American campuses concentrated on studying behaviour in the 1950s, Witelson learned the biological slant at Montreal's McGill University, taking notes from Donald Hebb, the man considered by many to be the father of "cognitive psychobiology," the Charles Darwin of his field. As chairman of the Psychology Department, the gaunt, bespectacled Hebb, a Maritimer who had been a farmer and a school principal before studying under Wilder Penfield at the age of thirty-two, encouraged his students to consider human thought in terms of the organ that actually produces it. Students who expected to explore Freud found themselves pondering the cellular substance of mind. Just as the tide began to turn away from psychoanalysis in 1949, Hebb published the influential book *The Organization of Behaviour,* which, among other things, uncannily described how neurons might effectively talk with one another, decades before research could prove him right. If the axon of one brain cell is close enough to excite another cell and fires it up repeatedly, Hebb theorized, some change takes place in both cells so that the power of that one cell to signal the other increases. Practice, in other words, makes perfect between neighbouring neurons. One of Hebb's most remarkable theories was that large groups of neurons firing together form a closed network of circuits he called "cellassemblies," and that networks triggered in a sequence actually produce "thinking."

During the time that Witelson spent in Hebb's department earning her doctorate, the buzz in neuroscience centred on the differences between the brain's left and right hemispheres. After intriguing discoveries were made in post-war epileptic patients who'd had their brains surgically halved, it became apparent that the left and right sides control different abilities. McGill scientists were among the first to design experiments into the so-called functional asymmetry of the two-sided brain, which set the stage for most of Witelson's future research. After she arrived at McMaster in 1969, she decided to study the interplay between the brain's two halves as part of a larger interest in children with dyslexia. She was recruiting boys at a Hamilton grade school when the girls said they wanted to be part of the experiment too. "The principal asked if I wouldn't mind testing girls as well," she said. "They wanted to get out of class and have some fun." The unexpected addition to her research project accounted for her first sex-difference discovery.

Witelson designed one of the tests to see how each side of the brain responds to touch and functions in the perceiving of shapes. The pupils had to reach behind a screen and in each hand feel a different, unusually shaped object at the same time for ten seconds. They were then asked to identify the shapes they had held in each hand from a set of drawings. Given that information from the left hand is usually processed by the right side of the brain and vice versa, and because the right side of the brain is thought to interpret information about shapes, Witelson assumed that the students would fare best at identifying objects in their left hands. But after testing two hundred boys and girls, this proved to be true—surprisingly —only for the boys. Witelson proposed that only in males does the right hemisphere of the brain do more work identifying shapes— or that somehow, in girls' brains, the left hemisphere communicates better with the right.

The experiment came to be called the Witelson test for assessing the spatial abilities of the brain's two hemispheres. The finding itself, meanwhile, led to her first report in *Science* in July 1976. Under the title "Sex and the single hemisphere," Witelson suggested that her research might explain why boys do not learn to read as quickly as girls, since one half of the male brain might not process the shapes of letters as well as the female's. Half a year later, *Science* published another Witelson paper that dealt with the role of the brain's right hemisphere in dyslexia, a reading disorder known to affect boys more often than girls. In the press, Witelson argued that boys should be taught differently from their female classmates. She predicted that boys would benefit from a "look and say" method of learning to read, to help integrate the verbal ability more controlled by the brain's left side and the spatial ability of the right side. Critics accused her of jumping the gun, arguing that there was no evidence that boys actually learned differently than girls. They suggested that feminism was unsettling science.

Despite the controversy, or perhaps because of it, Witelson gained a wide reputation as a leading expert in such matters. She had, after all, the distinction of being a Canadian who had won an American government contract to study the two-sided brain question. From 1976 to 1997, the U.S. National Institutes of Health (NIH) would eventually supply Witelson with more than $1.6 million to study "neuroanatomical asymmetry and psychology characteristics." In 1979, they sent an envoy to her McMaster lab to evaluate whether the money was being well spent. In an interesting coincidence, NIH officials asked Arnold Scheibel to fly up from California to do the site visit. So just a few years before he and Marian Diamond would take possession of those precious pieces in a mayonnaise jar, Scheibel gave the nod to keep funds flowing north for a researcher who seventeen years later would have her own chance to pick at Einstein's brain. "She had a small group of hard workers surrounding her, and I was impressed," Scheibel said of Witelson. "She was a very bright person and she had just begun her collection of normal brains."

✴

In the 1970s, medical oncologist Peter McCulloch was puzzled by his patients: specifically, the ones who didn't die. He'd assess their cancers of the stomach, bowel, lung or breast, their leukemias or lymphomas, and based on the life expectancy of others who'd suffered the same insidious scourges, he would reach a prognosis: a year to live, three years to live, six months. Patients hoped he was wrong; indeed, he hoped it. Sadly, his projections were usually right. Except that, every now and then, he and a couple of his colleagues at the Hamilton Regional Cancer Centre came across people who made a mockery of their predictions. "We'd have two patients, say, with the same type of cancer and give them two years to live; one of them would live two years and the other might live for ten," he said. "We wondered why people with the same disease had such vastly different outcomes." What advantage did they have? Was there an aspect of their immune systems that afforded them more protection than others? Was it genetic? Could their psychological makeup enable them to translate physiologically a will to live? As they mulled over the possibilities, McCulloch suggested that they drive up the road to McMaster and see if a psychologist would investigate the question. He'd heard Sandra Witelson's name in press reports and knew that she'd won a U.S.

research contract. He guessed that finding out whether the mind had some tangible impact on patients' bodies would suit her interests.

At the time, Witelson was wondering whether the mental differences between males and females might reflect anatomical differences in their brains. When McCulloch came calling she seized an opportunity. Witelson suggested that she could conduct neuropsychological profiles of the patients, interview them and test their mental abilities. Then, if the patients were willing, she could examine their brains after death. To have normal brains with corresponding profiles of the donors—including their IQ scores, whether they were left- or right-handed, writers, artists, musicians or math wizards— would be an invaluable resource.

McCulloch realized that Witelson had her own separate interests, but he thought that perhaps in the course of her research, she might find answers to his questions. Regardless, her plan sounded worthwhile to him. Dying patients often asked if there wasn't something altruistic they could do for medical science, since doctors had tried so hard to keep them alive as long as possible. McCulloch had to tell them no, for the most part. Medical schools generally aren't interested in using cadavers distorted by cancer for teaching anatomy. Nor could cancer patients donate blood, tissue or other organs, for fear of passing on the disease. But as far as anyone knows, cancer in other parts of the body does not affect the brain, so this would be different. "There was a real scientific basis for this," McCulloch said. "Nobody had done anything like this before: take mentally normal people and examine them and then study their brains afterward to correlate behaviour to anatomy."

Witelson refused most people who wanted to see her brain collection, which included catalogued and dissected tissue floating in formalin in a university cold room. In part, she feared it might be construed as macabre, the whole business of waiting for people to die so she could collect their brains. But the patients were going to die one way or the other, and as she once told author Robert

Pool: "The subjects get something out of it. They feel they're making a long-lasting contribution to mankind." Perhaps she also felt protective toward the brains, since she'd had a personal connection with most of the donors. McCulloch, whose work with Witelson never did uncover reasons for patients' outliving doctors' predictions, said, "As a clinical psychologist, Sandra was very good at dealing with the patients and their families."

She would have to be good for a patient running out of time to submit to twelve hours of tests that evaluated everything from general intelligence to personality, memory and even musical abilities. In a few cases, Witelson had discovered during the testing process that patients did suffer from mental conditions, and she would have to rule them out for her normal collection. Other times, patients eager to participate in the project discovered that their families were opposed. "You need to have the family's agreement. Once you die you don't own your own body, even if you've left it in writing," said McCulloch. "Sometimes the family gets leery, uncomfortable about doing anything with the body. They feel there's something sort of creepy about it." Doctors sometimes have to navigate convoluted family relationships when family members couldn't agree, or if the second wife of a dying man had not consulted the first wife. But once they had agreement, McCulloch said, the families worked with nurses on the project to get the body from the patient's house to an autopsy suite as soon as possible, often within three hours of death. They could be so determined to execute their loved ones' wishes that the nurse would "almost sit the corpse up in the front seat so they could drive them to the hospital themselves."

So, while most other neuroscientists battled the brain shortage, thanks to McCulloch's auspicious visit, Witelson never had to go without. Within a few years, she had enough brains to wade into a fascinating scientific debate.

After two American researchers reported in 1982 that a back portion of the corpus callosum, the arched inner bundle of nerve fibres that connects the brain's two halves, was larger in women than in men, other scientists had trouble reproducing the results. Witelson had the luxury of testing the question in forty-two post-mortem brains. If it were true, the implication of a bulkier female corpus callosum would add considerable weight to the argument that the two sides of a woman's brain talk to each other more than they do in men, and that the two sexes may indeed think differently.

As it turned out, Witelson found no difference in corpus callosum size between the sexes, but when she compared her measurements with the information collected from the cancer patients before they died, she discovered something else: left-handed or ambidextrous people had a significantly bigger corpus callosum than right-handers. Eleven percent larger, she estimated, which could represent as many as 25 million more nerve fibres. It raised the possibility that, as some suspect to be true of women, left-handers' brains work more holistically. Witelson's study landed on the cover of *Science* in 1985 and precipitated a string of discoveries related to the corpus callosum, handedness and the organization of brains. That sex got mixed up in it, as it had with her Hamilton grade school research, was inadvertent.

Witelson eventually discovered that the size difference in the corpus callosum between right- and left-handers applied only to men. Then she reported that there was a higher incidence of left-handedness among gays and lesbians than in the heterosexual population, suggesting that sexual orientation might have a biological basis linked to the way a brain is built. When she compared the mental abilities of gay men with those of heterosexual men and women, Witelson reached a controversial conclusion: gay men were in a league of their own. They did not, for example, fare as well as straight men on tests measuring spatial perceptions, but they did outperform the women. When they were asked to name as many animals as possible in ninety seconds, or draw copies of symbols within a fixed time limit, they did better than the heterosexual men, but not as well as the women. Homosexuals, her results showed, appeared to have an equal balance of spatial and fluency abilities—unlike heterosexual men and women.

It was the kind of scientific report that sends newspaper editors scrambling to find space in their next editions. Witelson told one reporter that her study "may help explain why there appears to be a preponderance of homosexuals within certain professions." She was careful to point out that she was not suggesting that one cognitive profile was better than the other, only that they were different. Some predicted that she would be accused of stereotyping gays, and even fuelling homophobia, and, to her irritation, she was. Not that it stifled her in any way. Two years later, in November 1994, Witelson reported that the corpus callosum tends to be bigger in gay men than in straight men, following from her earlier findings that a left-hander tends to have a larger corpus callosum and that more gay men tend to be left-handed.

Witelson published intriguing theories in the early 1990s linking her sensational discoveries. She referred to the idea that handedness generally reflects the criss-crossed organization of a brain in controlling the body and mind, and that there is a connection between handedness and sex. More males than females are, for example, left-handed. Drawing on other research into how sex hormones influence the construction of a brain budding in the womb, Witelson related it back to men and the corpus callosum. Since every fetus is thought to start out female, some think the fetal brain is feminine in design until it is exposed to male sex hormones. These hormones, she suggested, supervise the slaughter of a fetus's surplus neurons, destroying the axons that connect the two hemispheres in the corpus callosum of the brain. This, she thought, might explain why a right-handed man tends to have a smaller corpus callosum than a left-handed man. But lower levels of testosterone in the womb, perhaps

determined by genetics, might keep the neuron-pruning to a mini-
mum, which might in turn account for a larger corpus callosum and
left-handedness in men—and sometimes an attraction to the same
sex. To support her theory, Witelson noted that other researchers
had found that children born prematurely were more prone to suffer learning disorders, and were also more likely to be left-handed
than those born at full term. If preemies emerged from the womb
before sex hormones had a chance to sculpt a more svelte corpus callosum, it might explain their left-handedness, Witelson thought.

Left-handers have suffered a long history of persecution. In Latin
the word for "left" also gives us the English word "sinister"; in Italian
the word is *mancino,* also meaning "deceitful"; and the French word
gauche, for left, means "tacky" or "awkward" in English. Ancient tarot
cards feature a left-handed devil; seventeenth-century Salem witch-
hunters burned left-handers at the stake. Parents sure it was a sign
of bad character tied children's left hands behind their backs to force
them to conform to a right-handed world.

In neurology, the profile of a left-hander is one of extremes.
Although left-handers have the same basic brain design as right-
handers, left-handedness has been associated with dyslexia, epilepsy,
attention disorders and depression. Yet because both sides of their
brains tend to work in consort, left-handers are thought to have an
advantage in visual and spatial abilities. Some scientists suspect that
this is why left-handers, who in the West constitute about 12 percent of the population, work in the creative professions such as art,
music and writing in higher proportions. As well, some studies have
found left-handers overrepresented among those with a soaring IQ.

On lists of famous left-handers, Einstein's name frequently turns
up. Although in photographs he stands before a blackboard with the
chalk in his right hand and draws the bow across his violin with his
right hand, some claim that he performed other tasks with his left.
At the very least, he might have been ambidextrous. His name also

appears on lists of famous people born prematurely, along with Isaac Newton and Winston Churchill, who, like the brilliant physicist, were also reported to be late developers. Just where the notion that Einstein was born prematurely originates remains a mystery. Robert Schulmann has no explanation for it, though he is not at all surprised to learn that the century's great genius has become a patron saint of premature births, left-handers and even the learning disabled, offering hope that genius might be hidden in a handicap. "I heard from a friend in Germany that he is the crown prince of people with attention deficit disorder," Schulmann said. "That is complete crap." The young Einstein, after all, had his nose in Euclidean geometry at an age when other boys would be challenged to spell the words. He could focus so intently on his work that he forgot to eat, and sometimes sleep. Yet generations of students have tried to comfort themselves with myths about Einstein's ignoble intellectual beginnings. Posters and T-shirts celebrate his self-deprecating remarks: "Do not worry about your mathematical problems; I can assure you mine are still greater." Yet Schulmann points out, "There is no indication whatsoever Einstein was anything but bright in school." His teachers loathed him not because he lacked ability but because he was short on respect for authority. The only evidence that Einstein suffered from any disorder stems from anecdotes that as a child he was late to learn to speak. Some accounts have suggested that even at the age of nine he was not "fluent." Einstein himself said his parents were so worried about his speech that they consulted a doctor. When he did talk, he had the odd habit of repeating his words under his breath. But Schulmann said that even these reports might be exaggerated, since Einstein's maternal grandmother apparently cooed over his clever quips when Einstein was just two years old. "Where are the wheels?" he asked when his parents returned home with his baby sister instead of the toy he had expected. Yet all the traits ascribed to the rumpled genius, including his alleged premature birth and left-handedness—all of which Witelson heard—served only to raise questions about his brain: how big was Einstein's corpus callosum?

By the time Witelson told McCulloch that she had been asked to study Einstein's brain, her second thoughts had been shoved aside. "She was excited," McCulloch said. "She felt like it was an honour, really, that this guy came out of the blue to ask her to do it. She saw it as her being chosen as a result of her work." McCulloch was skeptical at the time that something meaningful could be found in an organ so long preserved. But if there *was* anything unusual, Witelson had a good chance of spotting it, he thought. She had an unrivalled control group of normal specimens from those unfortunate souls who never could have imagined their brains would be measured against the likes of Albert Einstein's.

✣

At night, from the Queen Elizabeth highway, Hamilton looks an ominous place. Rumbling up from the border across the Skyway Bridge you see the soot-spitting, corpulent beast of industry sprawled along Lake Ontario's north shore, belching into a black sky, twinkling its steel-mill lights across the harbour. But Harvey had seen all sides of Steeltown on his first trip: the verdant hills along Coote's Paradise and Hamilton Harbour, the limestone cliffs of the Niagara Escarpment rising out of the bedrock, slicing the city in two. Locals call it the mountain, and everyone in Hamilton lives on it or below it. Harvey checked in to the Mountainview Motel on Main Street. Witelson had recommended it. She told him that she and her husband Henry had stayed there when they had first moved to the city after finishing post-doctoral work in New York. The Mountainview is an old, low-slung three-star that offers a night's rest for the sweet bargain of forty dollars Canadian.

The next day it was show time. Not since the late 1950s had Harvey relished a morning like this, preparing to make a personal visit to a renowned scientist to share his prized specimen. In those days, fresh with expectation and pride, he'd had no inkling that the brain would become burden as much as bonanza. Now here he was at eighty-three, four years from the end of the century, in another country, hoping again for a solid study of the brain. He'd even brought his camera.

Harvey nosed the Dodge into the underground parking of McMaster's Medical Centre and left his things in the car. He'd need another set of hands to carry everything upstairs. Probably no one glanced twice at the pale, elderly gentleman as he limped on to the elevator. He and Witelson exchanged warm greetings in her office, a cluttered space of scientific texts, papers and a multicoloured plastic model of a brain. But Harvey wasn't ready to get comfortable. "Would you like to see the brain?" he asked.

He led her back the way he'd come, anticipating the mix of curiosity and thrill that streaked across people's faces the first time they saw it. Across the asphalt they walked to where he'd left his old Dodge. That's seen some miles, Witelson thought, spying the picnic on his front seat. Harvey popped the trunk and Witelson peered in. There, in the dank obscurity of a parking garage, nestled inside an old cardboard box, divided between two glass jars, were the uneven blocks of Einstein's brain.

In the lab, it was Witelson's longtime, silver-haired assistant, Debra Kigar, a master of crafting the plaster of Paris casts of the donor brains, who opened the jars and plunged her hands in to retrieve the pieces from their alcohol bath. Harvey had bundled some of the pieces in satchels of gauze to keep certain brain regions "together, so they wouldn't all be in separate bits." Witelson and Kigar both marvelled at the organ's fragments as they fell loose. "We held Einstein's brain in our hands and realized that this is the organ that was responsible for changing our perceptions of the universe," Witelson would later say, "and we were in awe."

As he had in Chicago when he visited Percival Bailey nearly a lifetime ago, Harvey recorded with his camera the moments that followed. But unlike Bailey, these scientists were not only ready to study
the brain; they were going to begin their research then and there.
The pictures convey a celebratory air as Kigar spread out more than
two hundred pieces of Einstein's brain, like a jigsaw puzzle, across a
large sheet-covered table in the lab. In another photo, Witelson beams
into the lens wearing an oversized Einstein T-shirt adorned with the
physicist's face and the caption "Velocity dilates time."

Harvey had brought all his photographs, which Witelson had
requested after learning that the organ was in pieces. Precise measurements would be virtually impossible to collect from a sectioned
specimen. If Einstein's fresh whole brain had withered by 135 grams
after two weeks in formalin, how much would it have shrunk after
its pieces had been encased in celloidin for forty-one years? There's
no room for compromise when a scientist measures neurological
structures in fractions of millimetres. For Witelson's purposes, the
calibrated photographs were as valuable as the brain itself. No other
scientist had ever had access to them. The others didn't know they
existed.

That first day, Witelson dove into Harvey's photo album and files,
admiring the calibrated images of the brain when it was whole. She
gained new respect for Harvey then, thumbing through the stack of
pictures and measurements and wondering where a clinical pathologist had found the resources to do all this in 1955. "He did a great job
of arranging all of these things that would have taken money," she said.
"I don't know what he did to get it; he didn't have a grant." Witelson,
meanwhile, would rely on the tens of thousands of dollars she received
from the National Institutes of Health and a grant from the Medical
Research Council of Canada for her work on Einstein's brain.

While Kigar arranged the jigsaw brain pieces in numerical order
to correspond with Harvey's map, with the old pathologist hovering

nearby, Witelson studied the photographs, preparing to document gross measurements of the organ. It was then, Harvey recalled, that Witelson stopped to analyze one of the pictures more closely. "Look at this," she said to Kigar. "Look at his parietal lobes." She was, Harvey gathered, impressed by the size of them.

✡

According to the ticks he penned on his maps, Harvey left Witelson fourteen pieces of Einstein's brain—more pieces than he'd ever left with anyone. Some are marked with her name while others simply say "Canada": numbers 74 to 76 from Einstein's right temporal lobe; 97 and 100 to 103 of his left temporal lobe; 20 from his right parietal lobe; 194, 195 and 214 to 216 from his left parietal lobe. Harvey told her that he'd sent slides and pieces to many scientists over the years, but that little fruitful research had flowed from his efforts. Witelson sympathized with him. She said scientists back then wouldn't have known what to look for. She said she was interested in tracking down some of those other pieces. At some point, while Harvey was there, she telephoned Sidney Schulman. The neuroscientist, who had retired from the University of Chicago three years earlier, had indeed built a large part of his career studying the thalamus of monkeys and other animals, just as he'd told Harvey he would back in the 1950s. "I was pretty surprised to get that call after all these years," Schulman said. "I told them I still had the slides of Einstein's brain. I keep them on my desk in a wooden box. It's tasteful; I don't show it to too many people. They wanted to know what slides I had and I told them they were from the thalamus. I said she could have them if she wanted them." Witelson thanked him and said if she needed them, she might get in touch.

Before he made tracks back to New Jersey, Harvey picked up his photo album from McMaster and felt a little uneasy. "Many of my photos were missing," he said. "I asked Witelson about it, and she said she would send them back to me when she was done."

One good look at the dimensions of Einstein's brain and Witelson sensed there was a heck of a lot there she might investigate. Thankfully, Harvey had had the foresight to take pictures in the midst of dissecting it, exposing the corpus callosum between the two hemispheres. Given that Einstein might have been ambidextrous, or as she termed it "not consistently right-handed," Witelson expected that the size of the structure might reveal something interesting, although she herself had reported back in 1991 that the corpus callosum—like the brain overall—tends to shrink as men get older, getting progressively smaller between the ages of twenty-five and seventy. Taking all this into account, she found that Einstein's corpus callosum consumed an area of 5.9 square centimetres, nearly a full square centimetre larger than predicted. But when she compared this measurement to her bank of brains—and for the Einstein study she used every available specimen in her collection, from thirty-five men and fifty-six women—she found his corpus callosum slightly smaller than the average. No matter. Witelson intended to compile and present all the measurements as part of a first full anatomical review of Einstein's remarkable organ. That very first day, she suspected that she had already glimpsed what set it apart. Not only did both the left and right parietal lobes appear unusually large to her—which was significant given the many studies linking this rear upper region of the brain to spatial and mathematical reasoning—but they also displayed an intriguing geographical pattern. When she later searched for a simple analogy, she would say it was as striking as seeing a face with the eyebrows beneath the eyes.

☼

In the seventeenth century, French anatomist François Sylvius was determined to draw a realistic picture of the brain's surface. Sloppy sketches before the Renaissance often featured the outer layer of human brains with no more definition than the topography of a walnut. Sylvius changed that. He compared the wrinkled cortex to the convolutions of a small intestine and launched the trend away from cartoonish depictions.

In the course of mapping the organ, Sylvius spotted a deep groove, a few millimetres wide, that sloped gently upward on both sides of the brain. It appeared to follow roughly the same route as the arms of eyeglasses. It began behind the root of the eyes and stretched backward to about the point of the temples. But instead of curling downward behind the ear as the arms of eyeglasses do, the crevice sloped up and backward and disappeared into a gorge-like depression just below the crown. To Sylvius it seemed a natural border; the cleft conveniently separated the front lobes of the brain from the temporal lobes at the side and the parietal lobes at the top. The "Sylvian fissure," as it became known, would serve as a major landmark for brain explorers to come.

None of this was unfamiliar terrain for Witelson. She and Kigar had closely studied the geography of Sylvian fissures. They had examined sixty-seven donor brains trying, among other things, to figure out if they could come up with a reliable way of measuring and describing the tricky route of the crevice. In 1992, they reported that "the Sylvian fissure anatomy in the human brain is very variable" and that no agreement exists as to its end point. But that was the peculiar thing about Einstein's brain, Witelson thought. His fissure seemed to take a unique detour in that its end point was nowhere near the crown. Instead of sloping upward and running into that gorge-like depression that scientists call a "parietal operculum," it stopped short. Witelson believed that the fissure in Einstein's brain was essentially stunted, that it melted into another fold. This pattern seemed most pronounced on the left side, where it left Einstein's lobe effectively undivided. Even more intriguing was that, to her, Einstein also appeared to be missing the parietal operculum on both sides of his brain. The absence of the gorge and the detoured Sylvian fissure seemed to account for the expanse of the great physicist's parietal lobes. Without these major cliffs and valleys interrupting the landscape, Witelson suspected that Einstein's neurons might have had an easier job of connecting with one another, in the way that her old professor, Donald Hebb, had once described. The theory seemed especially plausible given that the nerves on either side of the fissure and within the gorge itself appear to be involved in everything from movement to perception. As well, people are known to use their parietal lobes to process visual inputs, both real and imagined, perhaps the way Einstein envisioned himself astride a beam of light. "These are crucial centres for spatial visualization, for three-dimensional thinking, that certainly would be involved in the theories that Einstein was working with," Witelson said. "Perhaps the interconnectivity in that region is different, and maybe it conferred some advantage."

Above the Sylvian fissure in the frontal lobe, Witelson saw something else. It wasn't a groove that caught her attention this time, but a bulge, "a gyrus" or fold in Broca's area. Damage there will typically leave a person stammering or struggling to form sentences. Witelson wondered immediately if the anomaly in this area of Einstein's brain accounted for his alleged speaking difficulties as a young boy. But having not had the privilege of running the physicist through a battery of psych tests, as she had with the cancer patients, it would have been a leap to make the link between an anatomical curiosity in his brain and his real-life abilities, or lack thereof. The tenacious Witelson decided to settle for the next best thing. She would travel to Israel, where she could study the words written in Einstein's own hand at the Hebrew University archives in Jerusalem. An examination of his use of language, along with whatever biographical details she could assemble, might give her a feel for how the physicist thought. In short, she would try to compile a neuropsychological profile of the vaunted scientist forty-one years after his death.

Witelson might have completed her report sooner, but sadly, she had something else to write first. In February 1997, her husband Henry died. As a memorial for McGill University's Class of '63, Witelson wrote that her husband had suffered through a horrendous childhood. He had been born in Poland in 1937, two years before the Nazis invaded. His parents were murdered in the Warsaw Ghetto, and from the age of three he spent a lonely childhood in orphanages, jails and temporary homes. At seven, he was granted entry to Israel, and he lived with a foster family on a kibbutz until he was reunited with his older sister. They moved to Montreal in 1951, where, at seventeen, he met his future wife. Sandra Witelson wrote that Henry developed clinical depression in 1991, that the trauma of his early years had left its mark. "He continued working through these times," she wrote, "but the disease finally took his life."

So it was in an empty house, where she and Henry used to share takeout and a glass of wine before going to spend the evening working, that the widowed Witelson finalized the details of her anatomical study of Einstein's brain. McCulloch, who lived four doors down, saw her lights on in the small hours of the morning. "I think she overcame grief by working through it," he said.

☼

For Einstein, too, science was an antidote for sorrow. It was while mourning the death of his father in 1902 that he churned out his astonishing spate of revolutionary theories. Einstein once wrote that his father's passing was the deepest shock he had ever experienced. He had shared with his father his aspirations toward a career in physics, and Hermann Einstein had ached for his jobless son eking out a living on temporary calculation jobs and tutoring after graduating from the Zurich Polytechnic in 1900. According to the book *Einstein: Creator and Rebel* by Banesh Hoffman, a mathematician who worked with the Nobel laureate in the 1930s, Einstein's unsuccessful businessman father, despite being sick and a stranger to academia, pleaded with a prominent German physicist to help his son. "I beg you to excuse a father who dares to approach you, dear Professor, in the interests of his son," he wrote. "[Einstein] is exceedingly assiduous and industrious and is attached to his science with great love. My son is profoundly unhappy about his present joblessness, and every day the idea becomes more firmly implanted in him that he is a failure in his career and will not be able to find the way back again. . . . I permit myself to apply to you with the plea that you will read his article published in the Annalen der Physik and, hopefully, that you will write him a few lines of encouragement so that he may regain his joy in his life and his work. If, in addition, it should be possible for you to obtain for him a position as assistant, now or in the fall, my gratitude would be boundless. I beg again your forgiveness for my audacity in sending you this letter and I want to add that my son has no idea of this extraordinary step of mine."

Hermann Einstein insisted he be alone at his deathbed, which would always haunt his son, who, only a few years later, would never have to worry about finding work again.

☼

When the measurements and comparisons of more than two dozen brain parts were complete, the case, Witelson felt, was quite compelling. Stacked against the donations of the co-operative cancer patients, with a mean IQ score of 116 and sub-matched separately for age, Einstein's brain appeared to be built for brilliance. There seemed to her an uncanny symmetry to the cortex of the genius.

Each hemisphere of his parietal lobes bulged 2.5 centimetres wider than the control group, an average of 15 percent.

"Einstein's own description of his scientific thinking was that 'words do not seem to play a role,' but there is 'associative play' of 'more or less clear images' of a visual and muscular type,'" Witelson would recount in her report. It was the kind of thinking that might have sprung from the hind regions of his parietal lobes. She suspected that this wondrous advantage had not emerged from years of obsessively formulating theories and equations, but that it had sprouted long before then, perhaps when the brain was little more than a speck in the belly of a German woman. Early in his development, she would suggest, some chemical instruction had inflated the dimensions of his parietal lobes, effectively damming the normal fault line of the Sylvian fissure and the gorge it generally runs into and leaving a vast, open plane across which his neurons could forge their busy networks.

Witelson submitted the paper to the prestigious British medical journal *The Lancet* in February 1998. In it, she thanked her late husband Henry for his contribution. She included Harvey's pictures of the brain when it was whole and the measurements he had taken in those first weeks after the autopsy. Thomas Harvey and Debra Kigar were listed as co-authors. Witelson worded the article carefully and although it was entitled "The Exceptional Brain of Albert Einstein," she stressed that it should be taken as nothing more than a signpost, a hypothesis related to the biological features of intelligence. Nevertheless, sensation was bound to follow. After all, the McMaster neuropsychologist wrote that she could find no other brain embossed like Einstein's—not in the 182 hemispheres of her 91 control subjects, not in any specimen documented in the published collections of human brains. Her conclusions seemed to echo the words in which Harvey had always stored his faith: *someday we will learn things from this brain that we can't now.*

FIFTEEN

THE BIG BANG

☼

IN HIS EIGHTY-FIFTH year, tucked away in Cleora Wheatley's Titusville bungalow, the man who took Einstein's brain still inspired folklore of fantastic proportions. Although reporters had sniffed him out repeatedly since 1978, the facts, bizarre as they were, never managed to extinguish rumour. Thomas Harvey was discovered and rediscovered, like ancient ruins after a dust storm. This was in part because Harvey was a wanderer who travelled with his talisman, laying it to rest for a spell under a beer cooler in Wichita, at the office in Weston, under a pile of socks in Lawrence and then up there, on the hillside of Pleasant Valley Road. Ever recast against the backdrop of a fresh landscape, his was the never-ending story.

As late as 1997, Harvey's identity remained murky, even mythical. A young American magazine writer was enamoured of the tall tales he'd heard, and he had embellished them himself for effect over the years. Einstein's brain was supposedly stashed in a Saskatchewan garage gathering dust, and in the writer's imagination, the man who kept it had morphed into a hunchback with an eyepatch "pursued by secret agents or ex-lovers with an ax to grind." With these tantalizing bits of invention, and a suspicion that the truth might be stranger still, Michael Paterniti tracked Thomas Harvey to the treed slopes of southwest New Jersey. "I didn't know he was a writer when we first started meeting," Harvey said. "I thought he was a likable fellow who was interested to know about my relationship to Einstein." This was Harvey's euphemism describing the few minutes during which he'd drawn blood from the genius and the more than fifteen thousand days he had kept his brain.

Paterniti and Harvey met a few times, sipping tea in Cleora's basement, eating sushi in town. One night, at the Japanese restaurant, Harvey mentioned that he'd like to get out to California some time. Still buoyed by his trip to the Hamilton lab, and perhaps envisioning a replay at Berkeley, Harvey said he wanted to meet Marian Diamond. After all these years, he had a soft spot for the scientist who had first given his special specimen the serious attention he'd always thought it deserved. Down on his luck and between writing gigs, Paterniti offered to drive him out west, or, as Harvey put it, "to be my chauffeur." Harvey had a vague recollection that somehow it came up in their conversation that Evelyn Einstein also lived in California, near Berkeley, in the San Francisco area. As Harvey remembers it, Paterniti said he wanted to meet her. But Paterniti would later write that it was Harvey who suggested they visit Einstein's granddaughter. As with any writer and his subject, there are often conflicting recollections about the way events unfold. Paterniti assumed that the old pathologist "might be facing down some late-in-life desire to resolve the past with the Einstein family once and for all. Or, before his age permanently grounded him, maybe he wanted to hand the brain over to the next of kin."

Harvey selected a few pieces of Einstein's cortex for the trip. He wrapped them in gauze, placed them in an alcohol-filled Tupperware container and zipped it into a duffel bag. He'd been corresponding with a high school teacher in San Jose who had invited him to speak to her class. Harvey had given a few classroom talks back in Weston, and he thought maybe he could visit her school if he was out there. "But I'd never brought the brain with me in those other talks," Harvey said. "I only took the photographs. I just don't know why I took it [this time]. . . . I knew Marian Diamond had already received pieces of it, and I guess I thought I could show it to Evelyn. I remember thinking that Evelyn might like some."

With Harvey in the passenger seat of a rented Buick Skylark, Paterniti behind the wheel and Einstein's brain bobbing once again in a trunk, the unlikely couple set out in February 1997 on a cross-country adventure that would eventually read like an urban legend itself. With roadside diners and Las Vegas and the atomic history of Los Alamos, the journey offered a buffet of Einsteinian parallels and comic anecdotes for the artful wordsmith Paterniti. His travelogue, which first appeared in *Harper's* magazine, eventually became a book, *Driving Mr. Albert*, with Paramount Pictures snapping up the movie rights. "They said Paul Newman might play him," Cleora Wheatley said, after a movie agent had contacted them in Titusville. But Harvey didn't want to cede control of the way his story was depicted—even if it meant $100,000 in his own empty pockets. "They want me to sign something but I won't sign it," Harvey said. "They want to buy me off."

Harvey remains an enigma throughout Paterniti's telling, as mysterious as the brain itself, an old codger studying maps, imparting historical trivia ("Kansas is the biggest wheat producing state in the union"), yet content to pass wordless miles, no secrets revealed. The cargo they carry meanwhile distracts Paterniti. Like writers Steven Levy and Gina Maranto and so many others before him, he wanted to see the brain. Harvey had briefly shown it to him before they set out. But there, with the odometer spinning, Paterniti confesses: "I want Harvey snoring loudly as I unzip the duffel bag and reach my hands inside, and I want to—what?—touch Einstein's brain. I want to touch the brain. Yes, I've admitted it. I want to hold it, coddle it, measure its weight in my palm. . . . Does it feel like tofu, sea urchin, bologna? What exactly? And what does such a desire make me? One of the legion of relic freaks?" Paterniti does not lump Harvey in this category. But in his account the question lingers, as it always had: was Harvey a grave robber?

In the end, although Paterniti developed a fondness for the elderly doctor, Harvey said he was unhappy with the way the writer portrayed him. "I guess people think, you know, that there is some other story to tell just because so much time has gone by," Harvey said. The suggestions that he was unqualified for the job of minding Einstein's brain and that Princeton Hospital had fired him for refusing to give it up bothered him most. Paterniti quoted a neurologist who had worked under one of the esteemed researchers at the Washington meeting back in 1955 as saying, "Harvey didn't know his ass from his elbow from the brain."

It was Harvey's daughter Frances who typed his letter of complaint to *Harper's* from her home in Chevy Chase, Maryland. In it, the ever courteous Harvey thanked the magazine for financing his trip to California and said, "the whole trip was a pleasure and Mike's account of it is good reading though not exactly accurate." He tried to set the record straight, insisting that he had permission from Einstein's family to keep the brain, and that he left Princeton Hospital of his own accord. The last paragraph offers an accounting of his credentials and a glimpse at what was eating him. Referring to the nasty quote, Harvey said, the California neurologist "apparently thinks I know nothing about neurology. He did not know of my work under Zimmerman, nor under Dr. Fritz Lewey [sic]. . . . I taught Neuroanatomy for two years. . . ." Frances typed her own letter as well, defending her father as he would never defend himself. "While it is true that no one instructed Harvey to remove the brain before its journey to the crematorium," she wrote, "Nathan [whom she had met as a child] and the Einstein family expressed appreciation for Harvey's forethought, and ultimately granted him permission to act as custodian of the brain. Harvey simply assured the Einstein family that the specimen would be used only for scientific research. It appears that Harvey has kept to his promise. Albeit, Harvey's work remains unfinished, and stymies many curious folk who wish to have answers. This man's intentions clearly are not guided by a self-serving exploitation of the Great Scientist's popularity [that could be achieved at any time and in many ways]." Frances, like her siblings and friends like Lou Fishman and Cheryl Schimmel, worried that the world would be left with a horrible and unfair impression of Thomas Harvey.

☼

Evelyn Einstein had looked forward to laying her on eyes on the slippery fellow who had somehow made off with her grandfather's—possibly her father's—brain. In the nine years since Charles Boyd's unsuccessful attempt to extract DNA from it, Evelyn Einstein and Robert Schulmann had not made much progress in teasing out the truth of her lineage. In the meantime, she'd been plagued by health troubles. She lumbered under the strain of a weak heart and a liver condition that transformed every breath into an audibly taxing endeavour. She rarely left her apartment any more.

Evelyn had always assumed that Harvey was a thief. She never recalled her father saying anything about the brain particularly, but she suspected that if Harvey was given permission, it was by Nathan and not Hans Albert. "In all the articles printed after my grandfather's death, do you ever see my father quoted? Nathan always spoke for the family," she said. "Maybe Nathan didn't know how to get the brain back without publicity. . . . My theory is that [Harvey] was culpable and he damn well knows it."

When Harvey and Paterniti first arrived at her bayside apartment, "things were pleasant enough," she said. "Harvey told me he would send me articles on the brain and then he pulled out these Tupperware jobs and this box of slides . . . I felt disgusted, revolted and very curious."

Harvey apparently thought that Evelyn would appreciate Sandra Witelson's recent discovery, since, as Paterniti wrote, he began to talk slowly about "the fissure of Sylvius, the occipital lobe, the cingulate gyrus. All of it a part of some abstract painting, some hocuspocus act." Evelyn would say later that she felt Harvey became uncomfortable because she knew a thing or two about science and about brains from her anthropology days at Berkeley and Washington State. "I remember saying, I knew how much his brain weighed and that it was incredibly small, sort of microcephalic where 1920s brain research was concerned. All of a sudden, it was very strange, he just got up, bumbling and fumbling and picking up his things. He said he had to go and see relatives in the San Francisco peninsula. Then he asked to use the phone, and my apartment is small so I could hear him. It was clear these people didn't know who he was, he actually had to identify himself, he just wanted to bolt, to leave and go anywhere." As he'd left so many places and situations before that would have forced him to defend himself.

Memories conflict about what happened next. Harvey remembered that he asked Evelyn if she wouldn't care for a piece of her grandfather's brain, but said, "She didn't want any of it." Paterniti wrote that Harvey offered Evelyn a piece but then, oddly, never gave her one. Evelyn said Harvey never made the offer at all. Only after the old pathologist disappeared to see his relatives and she and Paterniti shared a meal did the two of them realize that Harvey had left the Tupperware in the backseat of the car, at last giving the writer and Einstein's granddaughter a chance to finger the damp remnants of genius. "Paterniti asked me, 'Don't you want to take a piece?'" Evelyn said. "I would have taken it if Harvey had offered a piece to me, but I didn't think that it was Paterniti's to give. . . . If there had been some kind of permission given for Harvey to keep the brain, then I suppose he had the legal claim on it technically . . . I would not have turned down a piece. I could have had it as an amulet for a

necklace, or a curiosity to take out to show people when one is feeling particularly ghoulish." Evelyn hoped to accompany Harvey when
he went the next day to visit Marian Diamond at Berkeley, but in the
end Harvey went without her.

✳

Since the publication of her paper in 1985, Marian Diamond had had
no contact with Thomas Harvey. And then there he was, riding up
the elevator, trundling down the hall past her lab, where nineteen
years earlier graduate students had tacked up an article about a remarkable brain resting in a cider box. "He's a hard one to figure out,"
Diamond said. "He was very reserved while he was here, a very polite
man." Harvey reached for his camera again and caught her there, smiling at him, a turquoise scarf tied round her neck. Her image would
be tucked away, alongside Percival Bailey's, in his old, well-creased
copy of *The Isocortex of Man*. Harvey asked Diamond whether she
would consider doing further study on Einstein's brain. She explained
that she "had gone on to other things."

Once a maverick, Diamond was now a respected authority in neuroanatomy circles. Her groundbreaking rat research had led to a suc-
cession of discoveries about the effects of environment on the
structure of a brain. She was also among the first to publish studies
on brain differences between men and women, and to demonstrate
why a person might never be too old to bolster the power of a cortex. The mental exercise of tackling crosswords and puzzles and mystery novels can help construct connections between neurons late in
life, she found. Scientists lent her research even greater credibility
when they reported the startling discovery in 1998 that neurons in
the human brain can continue to multiply late in life, refuting the
conventional wisdom that we are born with all the brain cells we
will ever have. Diamond went on to write three books, one of them

entitled *Enriching Heredity: The Impact of the Environment on the Brain*.
She was invited to lecture in Nairobi, China and Japan, where people
were surprised, she said, "to hear that a woman had looked at Einstein's
brain." Yet for all the accomplishments of her long career, none resonated more than the one that began with a mayonnaise jar. "Once I
published [my] study of Einstein's brain I just wanted to forget it,
but I couldn't because everyone is always calling about it," she said.
"People say you will be better known for Einstein's brain than all the
work that you did the rest of your life."

Had she been the type, Diamond might have gloated, knowing
that science had at last embraced her once outlandish theories on glial
cells. Several other research groups over the following years would
watch the supposed wallflower glia emit invisible signals, helping one
brain cell to connect to another, encouraging the growth of axons
and dendrites. Neurons working solo fail miserably in comparison.
Bruce Ransom no longer beamed Diamond's study on glial cells in
Einstein's brain onto lecture screens around the country to crack up
an audience. "They're not just nursemaids to the neurons; they're
partners, it's a true partnership," Ransom said. "I'm now absolutely
certain that glial cells are involved in learning and memory. . . .
Diamond gets the last laugh. She had an intuition, a prescience of the
most remarkable kind." While her report that Einstein had an abundance of glial cells in a wafer of his parietal lobe can never represent
anything more than a hypothesis, Ransom said, "She just needed more
Einsteins to study."

It amazed Diamond that Harvey sent her Christmas cards every
year after that visit. Harvey would explain that the gesture was
simple: of the scant published reports on Einstein's brain, he thought
that the Berkeley scientist did the best job. "Her paper is the most
important because it proved that a certain area of his brain was
significant and gave reasons for that," Harvey said. "Those reasons
were not too well known or understood at that time and she was

really a pioneer." But nostalgia also played a role in the way he ranked scientists' contributions. After all, Diamond had appeared out of the blue after he himself had effectively given up hope of hearing back from any of the scientists he'd canvassed for help. Diamond had put him and the brain on the map. While that turned out to be awkward where the media were concerned, it linked him to the scientific circles in which he'd been a stranger.

Considering the international splash it made, anyone might assume that it was in Sandra Witelson's report that Harvey found vindication. But Harvey would say he never needed validation of his meandering quest. Besides, soon after the Canadian researcher phoned to tell him about Einstein's unusual Sylvian fissure, frustration clouded any triumph he felt. He wanted his photographs back, the ones Witelson had taken from his album, the precious images of the brain whole that he had so carefully arranged with the Princeton University photographer in 1955. Twice Harvey called Witelson's McMaster office asking that she please return them. "I spoke to her secretary or her assistant but she never called back," he said. Even if she was continuing her analysis, Harvey wondered why she didn't simply make copies, why she needed to possess the originals. "I was trusting her to do what she said she would do and she didn't. She kept the photos." Witelson said that Harvey told her on the phone that he would like to visit her lab again and would simply take the photos back then.

To Harvey, the final insult appeared in ink. Even though Witelson explained in the text of her *Lancet* paper that the brain had been sectioned into 240 tissue blocks and that the pictures of it intact were taken in 1955, the caption beneath the photos reads: "taken in 1995 of five views of Einstein's whole brain." The typo, which *The Lancet* corrected one month after the paper was published, seemed to Harvey to make a mockery of the efforts he had expended, to compound all the accusations that he had never done anything with it.

"I think it makes her paper look like she did more work on it than she did," Harvey said. "Most people who read *The Lancet* wouldn't know that the brain was in pieces a long time before then."

☼

So much about science hinges on timing. Certain ideas popular in one decade are dismissed in another, so that science is often no more objective than any other human endeavour, vulnerable to the mood swings of the collective imagination, to the personalities and the desires of the people who drive it. Uncorked too early, the theories of an Oskar Vogt or a Marian Diamond spill like plonk. Wait too long and Freud is vinegar. Einstein himself passed nearly two decades in obscurity before the world applauded relativity. But the Witelson paper appeared three and a half years after Harvey's midnight crossing at Niagara Falls, on the cusp of a new millennium, when a study on Einstein's pickled brain was perfectly aged. A general optimism bubbled around all things scientific at the end of the twentieth century, much as it had in the postwar days when Harvey first took the brain, when atom bombs and antibiotics and the DNA discovery fanned a new faith in science. Now there was the wired world: the Internet on home computers and laptops, blinking cursors and the click of a mouse connecting faraway strangers and shrinking the planet. With the latest molecular rituals it seemed scientists could not only fathom nature; they could bend it at will: cross jellyfish with rodents and make mice that glow green in the dark; clone sheep and cows and create goats that shoot spider silk proteins from their downy udders; harvest stem cells from a human embryo and coax them to become muscle, bone tissue or heart cells that beat in unison; or crack the three-billion-lettered code that builds human beings. People banked on miracles. Parents saving for their children's education, greying baby boomers building retirement funds, grandparents padding dividends, all bought into promising technology stocks.

In 1999, science was NASDAQ hot, and though he'd been born 120 years earlier, so was Albert Einstein. The nostalgia-pumped media cranked out endless lists of the century's best this or most that, and his name always hovered near the top. Pundits said he embodied the scientific spirit of the age; he was an ambassador whose theories laid the groundwork for the century's great achievements: electronics and nuclear power, space travel and television. It is hard to imagine a more auspicious moment for a scientist to have sallied forth with a secret from Einstein's brain.

After the text had been laid out on the page, draft copies proofread and corrections faxed back and forth, *The Lancet* kicked its public relations machine into high gear. Unlike *Experimental Neurology*, where Diamond published her work, or *Neuroscience Letters*, which contained the Anderson report, the prestigious *Lancet* is a vintage British medical journal, founded in 1823 and internationally renowned. Its editors know a sizzling story when they see it. Advance notice of the Witelson paper flickered onto computer screens in news outlets around the globe. For big journals like *Science* or *Nature,* it's customary to distribute early, electronic versions of their contents to give reporters enough time to gather facts and commentary to write a more complete story for a public hungry for science news. That kind of media attention also gives the journal free publicity, bolstering its stature because the lay press deems its content newsworthy. Sandra Witelson herself realized that her paper was bound for the headlines. Though researchers are often reluctant to trumpet their work, even the high circles of academia hope to cash in on the value of good press. Politicians who control the purse strings of granting agencies don't read science periodicals; they read the papers. Witelson's office took the bold step of actually phoning health editor Paul Taylor at the *Globe and Mail* a few days before the *Lancet* publication to tell him that they had discovered something interesting in Einstein's brain. The catch is that writers who receive a report early must honour an embargo, agreeing to wait and publish their accounts around the same day the journal does or risk losing future advance notice. The arrangement offers researchers and the journals the chance of a "big pop" or a "blitzkrieg," as they call it in the public relations world, where a story appears in several places all at once, saturating the nightly news, the newspapers, the radio, magazines.

☼

The sky hung grey and bloated over Hamilton on June 17, 1999, cooling the sticky air and threatening rain. Witelson was in her office, glued to the phone, a stack of messages before her from media anxious for five minutes of her time. "How did it feel," they wanted to know, "to touch Einstein's brain?" What did she think it meant, this business about his Sylvian fissure? In the public relations office, the lines had been busy all week. Out in the reception area, behind a fabric-covered desk divider, a weary office assistant sighed and rubbed her forehead. "It's been crazy in here, just crazy," she said, snatching up the receiver of the ringing phone. "Okay, I'll take your number and Dr. Witelson will try to call you back . . . soon . . . I can't say when."

ABC and CNN had called, the assistant said. They were considering a live hookup and the *Hamilton Spectator* was sending a photographer that afternoon. Witelson called out from her office, asking if anyone had heard back from *The Lancet* on a last-minute query. Then she popped out, her curls clipped back on one side with a rhinestone butterfly, *The Lancet* paper drooping in her hand, pointing and nodding her approval to an earlier correction. It was nearly three o'clock.

"Have you had lunch?" her assistant asked.

She shook her head. "I haven't had a minute today," she said apologetically to a waiting reporter. Then she whipped back into her office, slipped on a smart black leather blazer and in a flurry of parting instructions and a promise to return soon, she dashed down to the cafeteria for a cheese sandwich and an interview, telling the story, as she would innumerable times in the days to come, about a curious fax that rolled into her office.

It might be said that for every scientist there is a career-defining moment. It arrived for Frederick Banting and Charles Best in 1921 with the insulin-cured dogs in a University of Toronto lab. For British doctors Patric Steptoe and Robert Edwards it came on a July day in 1978 when Mrs. Lesley Brown delivered a test-tube-conceived baby girl. Einstein felt it after a balmy night's sleep, when he turned to his friend and said, "I've completely solved the problem." Harvey experienced it that spring day in the morgue, a blade whirring in his hands. For Witelson it might very well have come on June 19, 1999, the day the report on Einstein's brain finally hit the press, a moment that crossed the career peaks of those two other men, strangers really, colliding in cosmic flashes of space and time.

News that neuroscientists had sussed out a possible explanation for Einstein's brilliance turned up everywhere, from the front page of *The Plain Dealer* in Cleveland, Ohio, to the Mexican paper *Reforma*, from the *Lethbridge Herald* to the *Times* of London, from the *Globe and Mail* to *La Recherche*. It appeared on the CBC, the BBC, ABC and a veritable alphabet of networks too long to list. The university public relations office quit keeping tabs on all the programs that aired the story. Despite the institution's long history of accomplishments—finding how an old heart drug could save thousands of lives, discovering how babies learn to see, pioneering keyhole cardiac surgery—university officials were unequivocal: the unusual anatomy of Einstein's brain, the campus newsletter said, was "the biggest media story McMaster has ever seen."

On top of all the scientific hullabaloo and *fin-de-siècle* resurrection of Einsteinmania, the Witelson report on the brain of all brains also happened to cap off what former U.S. President George Bush had dubbed the "Decade of the Brain." The biology of behaviour—or "cognitive neuroscience" as its practitioners call it—had exploded. Universities could not fill jobs fast enough. The mix of a psychologist with an intimate understanding of grey matter was all too rare. Colourful, moving pictures of the brain in action generated much of the excitement. Scientists no longer had to wait for a patient to bash his head or have a stroke to deduce what function a particular area of the brain might control. Nor did they have to electrically tickle the organ's whorls in a living subject to record the reaction it produced, as Wilder Penfield had once done. Armed with multi-million-dollar scanning equipment, they could watch the brain in action. Subjects slid into the inner tubes of the functional magnetic resonance imaging machines. Scientists meanwhile surveyed the magnet-tracked blood flow through their brains as it ferried oxygen and sugar to the areas believed to be busy thinking. They could witness blobs of colour materializing on a screen and pinpoint the active machinery of a mind spelling, crying, speaking a second language. They could see the right frontal lobe process a joke, the almond-shaped amygdala recognize a picture of Mother, see the parietal lobes light up like a pinball machine when solving math problems or guessing the number of dots on a line. Had Einstein been alive, might they have seen sharp flashes in that same space above his ears as he straddled his imaginary beam of light? Thrilled with the speed of their progress, scientists were filled with grand hopes. Witelson herself became fond of saying that 90 percent of what's known about the brain they had learned in the last fifteen years. Neuropsychologist Donald Stuss, director of the Rotman Research Institute in Toronto, predicted that scientists would, in the near future, "be able to describe the unfolding of a thought millisecond by millisecond." At the beginning of the Decade of the Brain, Israel Lederhendler, the chief of behavioural neuroscience for the U.S. National Institute of Mental Health, described it as an exhilarating sense that "things are about to be understood."

Amid this high-technology fervour, the irony was that in analyzing Einstein's brain, Witelson had required tools no more sophisticated than forty-one-year-old photographs of it, calipers and a measuring tape. Rudolph Wagner had used a similar method in 1860 to study the post-mortem brain of Karl Friedrich Gauss. Wagner too sensed that he'd spotted a clue to Gauss's mathematical wizardry in his heavily wrinkled cortex. Witelson actually referenced this first known biological hunt for genius in her own paper, noting that the brains of both Gauss and the physicist P. A. Siljestrom had shown "an extensive development of the inferior parietal regions." She made the case that such turn-of-the-century studies of great brains had been hamstrung, since researchers usually knew nothing about the mental state of their subjects when they died. Neither did they have a group of normal brains to compare them to, "so that the results were mainly idiosyncratic observations." What's more, she wrote, they had no idea which particular brain areas might control a specific activity and so many treated intelligence as though it were a singular trait, ignoring the fact that a dazzling mathematician might also be a lousy poet. So they weighed brains, assuming that bigger meant smarter, only to become discouraged when the likes of French Nobel laureate and writer Anatole France rung it at a measly 1.017 kilograms. And now here was Witelson, at the turn of the next century, with the heft of history and the vibrant images of scanning machines behind her, making the case that size might matter after all. The theory offered no end of quips for headline writers: "Revealed: The big secret of why Einstein was so brainy," said the United Kingdom's *Express*; "Einstein was bigger where it counts," declared the *Sydney Morning Herald*; "Insights into a Mental Giant," read *Maclean's*.

When Witelson flew to the Society of Neuroscience conference in Miami in the fall of 1999 and gave a talk on Einstein's brain, no one shuddered at her linking the size of the physicist's parietal lobes to his particular brand of genius. Oskar Vogt's audiences might have

stifled their laughter when he described the impressive girth of an
auditory cortex in a violinist with perfect pitch, but no longer.
Scientists surrounded Witelson after she spoke. They asked for details,
perhaps mildly curious, too, about her experience studying the brain
of one of the greatest scientists who ever lived. Tomas Paus, a neuropsychologist from the Montreal Neurological Institute, was among
them. "Her research was solid," Paus said. "Whether we like it or not,
there seems to be a correlation between capacity to perform com-
putations and the size of a parietal lobe."

That same fall, Paus had reviewed a fascinating paper that would
soon make news itself. Neurologists at University College, London,
concluded that the brains of the city's famed cab drivers might actually bloat with "the Knowledge," a nickname for the encyclopedic
information of some seventeen thousand streets and ten thousand
geographic locations that drivers must memorize before landing a
licence to taxi people through England's capital. Compared to non-
drivers, the cabbies possessed a considerably bigger hippocampus,
the inner brain's amazing sea-horse-shaped library of permanent
memories and mental collections of maps—the quickest route, say,
from the Royal Society for Medicine to the Royal Hospital for Tropical
Diseases on St. Pancras Way. Researchers also found the longer a
cabbie had driven, the bulkier the hippocampus became. "Size is
coming back, more and more," Paus said, "even though it has been
controversial."

In 1995, German researchers reported that an MRI study showed
that musicians who had begun their training before the age of seven
boasted a larger corpus callosum, with apparently more nerves connecting the two halves of their brains. In 1996, Baltimore scientists
relied on scans to conclude that boys' brains are on average 10 percent larger in volume than those of girls before the age of five and
that there is a link between greater volume and a higher IQ. By
1998, as many as eight different groups reported MRI results that

suggested a statistically significant connection between larger brain size and a higher IQ. A month after Witelson's paper appeared, researchers in Pittsburgh reported that they had actually measured the heads of mentally healthy, elderly adults to investigate whether size had any bearing on how the subjects performed on a cognitive screening test. The study's title summed up their findings: "Small head size is related to low Mini-Mental State Examinations in a community sample of nondemented older adults." (Yet the taboos still applied. Those who wanted to report on brain-size differences among racial groups had a hard time finding journals willing to print their work. Australian researcher Clive Harper, for example, who found that Aborigines have a visual cortex 25 percent larger than Caucasians, imbuing the natives with an uncanny ability to recall the precise location of a waterhole in the parched expanse of the Outback, was refused permission to even outline his evidence at an American conference.)

While the brain-size trend reeks of conceptions tossed out with crinolines, Paus said that the core question now is not "how big is the brain?" but "how is the brain big?" The enlarged brains of people with Down's syndrome, for example, reflect a foul-up that began in the womb: instead of the usual pruning of neurons with the mass suicide of budding brain cells, the Down's syndrome brain is inflated and crippled by their surplus survival. A tumour can swell a brain while doing nothing more than perhaps harming intellect. And measuring the circumference of a whole head probably tells you nothing. According to Paus, "This whole issue is a minefield and one has to be very careful. You could be accused of phrenology of the twentieth century, measuring the size of the bumps on the cortex instead of the skull. So you have to ask things like whether a gene caused a particular part of the brain to grow more neurons or more glial cells. You have to ask what makes it bigger." Paus guessed that the London cab drivers might not have more neurons in their hippocampi, but that their neurons are busier and require the support of more glial cells. If the average ratio in the brain is ten glial cells to one neuron, a person might have only fifty glial cells if only five neurons are active. But if ten neurons in the hippocampus are firing full blast, one hundred glial cells might spring up to nurture them, he said, expanding the overall mass of tissue. Paus said the same theory might explain why Witelson found that Einstein had a parietal lobe 15 percent wider on both sides than those in her control group. As Diamond had suggested, to much ridicule fourteen years earlier, Einstein might have had more glial cells.

Under the headline "His Brain Measured Up," Canadian-born neuropsychologist Steven Pinker, a professor at the Massachusetts Institute of Technology and one of the field's most eloquent spokesmen, wrote in the *New York Times* that the Witelson report was an "elegant study . . . consistent with the themes of modern cognitive neuroscience." He found that the unusual route of Einstein's Sylvian fissure was not a subtle difference, and that the finding was particularly intriguing given the presumed functions of the parietal lobes: "It is strangely fitting that the brain that unified the fundamental categories of existence—space and time, matter and energy, gravity and motion—should now be helping us unify the last great dichotomy in the conceptual cosmos, matter and mind."

Not all reactions to Witelson's work were as positive as Pinker's. Once again, as with Diamond's report and also Britt Anderson's, the study reignited the issue of the long-serving brain and Harvey's guardianship of it. Among those who read *The Lancet* article was a young Baltimore endocrinologist bothered not so much by its conclusions but by the mere fact of its existence. It put Roberto Salvatori in mind of old Harry Zimmerman, lying awake in a hospital bed in the Bronx, recounting his own experience with the brain of the genius. In the years since Salvatori had trained in New York, he'd launched an impressive research career at Johns Hopkins studying the genetic causes of dwarfism. But he had never forgotten that chat on the overnight shift, and he assumed that the Witelson report would have disappointed Zimmerman—did it not defy Einstein's own wishes? "It saddened me to see the article," Salvatori said. "Based on what Dr. Zimmerman told me, I don't think it's what Einstein would have wanted."

"Sir, I am [a] little troubled by the publication of Witelson and colleagues' report," Salvatori wrote to *The Lancet*. "When I was a medicine resident at Montefiore Medical Center in New York (University Hospital for the Albert Einstein College of Medicine), I had the honour of taking care of the late Harry Zimmerman, a famous neuropathologist, who was one of the founders of the medical college. Zimmerman told me the tale of how difficult it had been to convince Albert Einstein to allow his name to be used for the medical college. He also told me that he had obtained from Einstein permission to examine his brain after his death in search of some anatomical marker of exceptionality. According to Zimmerman (who told me that Einstein's brain was absolutely normal), Einstein's permission was conditioned on his promise not to publish the results of such a necropsy."

Witelson defended herself with Harvey's version of events, writing that "consent to study Einstein's brain was given to Harvey in 1955 by Einstein's elder son, Hans Albert Einstein and by Nathan, with the provision that the results be published in scientific or medical journals." This was enough for Witelson. Permission from family members, in her experience, counted as much as the consent of the cancer patients themselves when determining whether she could add their brains to her valuable collection. Regardless of what the patient wanted, the relatives had the last word. Family members could contest any instruction the deceased left in the will. She and the nurses running the brain collection project always made a point of reaching out to them.

One scientist questioned whether Witelson had actually studied Einstein's brain at all. In another letter to *The Lancet*, Albert Galaburda, a Harvard Medical School neurologist with the Beth Israel Deaconess Medical Center, wrote, "We are given no information to ascertain that the photos indeed came from Einstein's brain, which is not a trivial issue after decades of obscurity." He wondered, too, what really might be said about the state of Einstein's brain at the time of his death: had he fallen unconscious in any way from lack of oxygen, damaging the organ even before his aorta ruptured and his body quit?

Beyond that, Galaburda found fault with the study on substance and principle. Back when Diamond had completed her study, he'd criticized the effort in a Boston newspaper. Reading the Witelson report ruffled him again. "I think it's tempting to find some easy answer to a very complicated question. I know these are respected researchers, I know both these women [Diamond and Witelson] and they're not stupid, they are very intelligent women, but I don't know why they do these things. They are seduced." Still, Galaburda said he himself wouldn't turn down the chance to look at Einstein's brain. But he would never do a study, he said. "It really bothers me to see this kind of thing come up again, it irritates me. . . . I'm not saying there isn't a secret there but that it has so far defied discovery and I'm not saying I don't think it's worth trying." In 150 years of its "yucky" history, Galaburda said, no study had successfully correlated intelligence to the shape of the brain, and Witelson's take on the Einstein organ was no exception. "It's ridiculous to claim the pattern of Einstein's Sylvian Fissure was unique." To his eye, the deep cleavage that divides the parietal lobes was often asymmetrical in right-handed people, which he argued that Einstein was. As well, Galaburda felt that he could actually see from the photographs published in the journal the gorge, or parietal operculum, that Witelson believed was missing. "I think he's got a very typical pattern for right handed people. . . . I don't think these people know how to look at a brain," he said. "They were mislabeling things, they were making huge leaps of faith." In his letter, he referred to a classic 1953 text on the parietal lobes and said that one of the photos presented in it "is strikingly similar to the figure presented as Einstein's left parietal lobe . . . Einstein's brain is no exception to the most common of patterns."

When Harvey first heard that Witelson thought Einstein had an unusual Sylvian fissure, it didn't entirely surprise him. "I remembered thinking at the time [after the autopsy] that it looked short," he said. But the Princeton pathologist could attach no significance to it. As his bible, *The Isocortex of Man*, told him, there is a "bewildering variety of sulcal patterns in the parietal lobe reflecting that the brain has a particularly high degree of 'developmental freedom' . . . The brain is the most individualistic of all organs."

Jay Seitz, a psychology professor at City University in New York, wrote to *The Lancet* on June 22, 1999, saying, "In a study that purports to look at the extraordinariness of individuals, it is odd that the control group consists of only one . . ." He argued that Einstein's brain should have been compared to a collection of other physicists and mathematicians and other groups of talented individuals. He suggested that if Witelson could not track down such post-mortem brains, she might have used brain imaging for the study, as she once had in measuring the corpus callosum of gay men.

Like Seitz, neuropsychologist Doreen Kimura, who had trained under Donald Hebb at McGill, felt the work of her fellow alumna was an "initial observation [that] should have been the beginning of a proper study, not presented as a finished product. As to why *Lancet* published it, I suppose the same reason a lot of people got excited about it. It's a simplistic explanation of something. Lots of bad science gets published even in prestige science journals. . . . But the public loves a simple story, however misguided." Kimura, a professor at Simon Fraser University, pointed out that Einstein's brain was compared to the average measurements of a large group of "normal" people in the *Lancet* paper. But was Einstein's configuration never found in any of the ninety-one other subjects? How many of the ninety-one people had an inferior parietal area that was 15 percent wider than "normal"? Was Einstein alone in this extreme? "You can imagine, since brain features vary across individuals, and in some sense each brain is unique, that you could have the following scenario: a very good carpenter dies and his brain is donated for autopsy. He has a largish area in one part of the brain. Someone therefore concludes that this is where his carpentry skills reside. Would we accept that conclusion without further information and other [carpenters'] brains to compare? Hardly."

But no one would publish a report on the brain of an anonymous carpenter. Critics argued that a study on Einstein's brain takes on a significance that no other single-specimen study would, simply because it adds another detail to the endlessly fascinating story of a very famous man, much the way that a celebrity suffering from an illness gives it a public face. Lou Gehrig is as well known for the nerve disease that killed him as for the brilliant baseball career ended by amyotrophic lateral sclerosis. Modern-day examples are endless: *Superman* star Christopher Reeve speaks for spinal cord injury; Michael J. Fox and Muhammad Ali for Parkinson's disease. Having a marketable personality to act as spokesman is a tool so effective that the University of Rochester's savvy PR department actually promoted recent multiple sclerosis research with a reference to the fictional president Jeb Bartlett, played by Martin Sheen, who battles the neurodegenerative disorder on the network hit *The West Wing*. As Frederick E. Lepore, a clinical neurologist who lives three blocks away from Einstein's former house in Princeton, put it: "For better or worse, our culture readily accepts the lessons taught by afflictions of important brains. It is hardly surprising that critical scrutiny may be suspended when secular priests in the white raiment of science reveal that Einstein's brain is indeed different."

✷

Witelson stood by her findings. She acknowledged that her study would have benefited from access to other gifted brains for comparison to Einstein's and that her report "clearly does not resolve the long-standing issue of the neuroanatomical substrate of intelligence." But she spent little time fretting over what the critics had to say. Within a week of the paper's publication, she was back in the news. McMaster announced that Hamilton businessman and philanthropist Irving Zucker, well known for his generosity in supporting the city's art gallery, theatre and opera house, would give Witelson $1 million for research. McMaster in turn matched the money with an additional million and named Witelson to the first Albert Einstein–Irving Zucker Chair in Neuroscience. The seventy-eight-year-old Zucker told the *Hamilton Spectator* that Witelson had "worked on me for three years. Now I know why she's so successful. She goes in there and digs and digs and digs." So many attended the Zucker announcement that the event had to be moved out to the campus lawns. Certainly her earlier work had attracted its own share of public attention, but Witelson had never been on the receiving end of anything like this. In addition to the television interviews and talk shows, educational leaders and business types paid to hear her speak. In the company of Spike Lee and Silken Laumann, Magic Johnson and Stephen Jay Gould, Peter Gzowski and, coincidentally, Gregory Stock, Sandra Witelson registered with the David Lavin Agency to handle her $3,500-a-pop speaking engagements.

Four months after her name had been inextricably linked with Albert Einstein's, Witelson sipped a Cabernet at a Hamilton bistro, her charcoal eyes set off by an avocado V-neck sweater. She didn't know how celebrity would change her life, she said. It had been fun. From her satchel she tugged the August 30 issue of *People* magazine, the celebrity industry's weekly scripture. A smiling Julia Roberts and Richard Gere, 1999 nominees for the sexiest examples of their genders, adorn the cover promoting their latest movie with an inset photo of a real-life "runaway bride." Witelson's small, manicured hands flipped through its pages. And there, wedged between stories about novelist Danielle Steel divorcing her fifth husband and the forty-year-old cousin of the late John F. Kennedy Jr. losing a battle with cancer, is the McMaster neuropsychologist smashing the stereotype of a white-coated scientist. Robed in black lambskin, softly lit and with her elbow draped over a plastic model of a brain, she is quoted in the caption beneath: "Can you imagine the feeling? That is Einstein's brain in our lab." *People* magazine sent someone all the way from Chicago to interview her, she said. And they sent a photographer to her house, which she usually doesn't allow. She's had a handsome book offer to tell her story and, she said, even an invitation to co-chair a world conference with Nelson Mandela. She had no idea, she said again, how this would change her life.

By year's end, as *Time* magazine declared Einstein its "person of the century," Witelson's involvement with the dead genius spawned its own lore and legends, just as Einstein's vaunted brain and its keeper had done. Witelson's celebrity seemed to have less to do with her study than it did with the mere fact that she possessed a precious, physical piece of the Man of the Millennium. Witelson never talked about where or how she kept the tissue. "I don't think that's relevant," she would bristle. Yet the pickled tissue stood as the essential ingredient of her fame. Canadian media tycoon Moses Znaimer, the hip force behind Toronto's City tv, included Witelson on the list of elite presenters in 2000 when he organized the country's first TEDCity conference (TED standing for technology, entertainment and design). Among impressive names in music, fine art, literature, business and science, the official website for TEDCity 2000 described the McMaster scientist this way: "Dr. Sandra Witelson has not only conducted the first full anatomical study of Albert Einstein's brain, she is also still in possession of it." An arts report in the *Globe and Mail* leading up to the June event read: "Fork over $3,000 bucks to be a delegate, and meet Sandra Witelson, a neuroscientist at Hamilton's McMaster University who just happens to own Albert

Einstein's brain (legend has it, it was found in a pickle jar in some-one's car and she bought it)." When the *National Post* covered Witelson's talk at the same conference of "cultural elites" a year later, the misperceptions had become entrenched. The *Post* identified the McMaster scientist as the "owner of Einstein's brain" in its head-line, reporting that "Harvey sought out and selected Witelson as a suitable guardian for this anatomical artifact." The article went on to explain that "Witelson—and through her the rest of the world—has learned much about how the brain works by studying Einstein's brain." "I still have it," she told the newspaper. "I won't say where it is but it's in a place no one will find it." That she had, according to Harvey's maps, only fourteen pieces, roughly one-fifth of the cov-eted organ, was a point that the article did not clarify. Witelson wanted more pieces to conduct further studies, but that seemed unlikely. Harvey said he told her he would send no more tissue unless she returned his photographs. But Witelson had discovered that the elderly pathologist was no longer the person to ask.

✡

Some say the universe started with a single explosion, propelling dense matter far and wide with the force of its own burst scattering shimmering particles into the expanding darkness of space. The cult of celebrity has become a similar cosmic phenomenon. It expands and engulfs, feeding the vacuum of cable airspace with everything from the marvellous to the mundane. Tell-all talk shows feature teary-eyed confessors, and entertainment programs report even the most arcane tidbits from the entertainment industry, offering every atom and neutrino its fleeting moment of fame. California State University media psychologist Stuart Fischoff actually calls it the Big Bang the-ory of celebrity. In the midst of the eruption, people can become famous not for swimming across Lake Ontario or flying solo over the Atlantic, authoring the theory of relativity or achieving some other wondrous feat, but merely for crossing the path of some other dazzling meteor. Fame, like stardust, rubs off. As Einstein himself said after visiting celebrity-obsessed America for the first time in 1921: "I really cannot understand why I have been made into a kind of idol. I suppose it is just as incomprehensible as why an avalanche should be triggered off by one particular particle of dust, and why it should take a certain course."

Fischoff's ideas grew out of cases like that of Kato Kaelin, an aspiring actor and sometime babysitter who was catapulted from obscurity to wealth and fame simply because he was a guest at the O. J. Simpson mansion the night Nicole Simpson and Ron Goldman were stabbed to death. In *People* magazine, an unknown woman who walked out on her wedding can find herself sharing the cover with grinning movie stars, and a young man's battle with cancer takes on greater significance because he happens to be the cousin of John F. Kennedy Jr. The concept was not foreign to Evelyn Einstein, who often found herself snickering at official functions commemorating her grandfather as she eavesdropped on guests claiming to be distant family members connected to the genius by marriage or a cousin twice removed. "I had no idea I had so many relatives," she would laugh. As Fischoff explains it, "Anyone connected to the penumbra of celebrity gets sucked into it. People are famous by virtue of their connections, if you can't have the celebrity . . . you talk to the person who did her laundry. . . . If you can't have Albert, you can have the next best thing, the woman who has his brain, and people will expect somehow to be enlightened." Einstein himself once had an opportunity to share the spotlight of another celebrity. When the physicist accompanied Charlie Chaplin to the premiere of his film *City Lights* and throngs of adoring fans mobbed their limousine, Einstein asked the silent screen star, "What does it mean?"

Chaplin replied, "Nothing."

✳

Down in Birmingham, Britt Anderson saw the August issue of *People* and Sandra Witelson gazing out from its glossy page. "Whew," he thought. "I guess she maybe has a good PR machine, or it could be that she published in a prestigious journal. I just had this loud silence." He'd read the Witelson study when it came out and looked to see whether she referenced him. "There were only two other published studies on [the brain], after all." Anderson still included his report on Einstein's brain on his curriculum vitae, and it invariably elicited curious questions: "Do you really have the pieces? Do you rub it for luck?" He had tried other experiments with silver stains on the tissue to test whether he could see signs of aging in Einstein's brain. Plaques generally show up under the microscope as reddish circles or thick, brown squiggly lines that mark the spot of a damaged or dead neuron like a tombstone. But once again, celloidin thwarted his efforts. He hadn't examined Einstein's brain since. "I don't have any plans to do anything further with it," he said. "I have moved further along in my interests in science and I've lost the lustre of being in the company of this tissue. I have also become circumspect about the purity of my own motives for wanting to do this. What really is the scientific value? What can the study of this one brain tell us?" he said. Even when his paper was accepted for publication, Anderson thought that, while it was provocative, it was really nothing more, maybe not even a starting point. He came to think that other scientists who studied it were trying to be opportunistic, "like I was trying to be opportunistic. . . . If Witelson wanted more I'd probably send her my pieces."

Anderson has an entirely different plan now to capitalize on his experience with Einstein's brain. With a wife and two children and his fortieth birthday just behind him, he's planning to quit his job, give up his income and go back to school to learn how to build three-dimensional computer models of the brain so he might be better equipped to solve the mystery of intelligence. He keeps a motto tacked to his office bulletin board that Harvey would appreciate: "Making a living is satisfying enough but sometimes you have to make a difference."

Geneticist Charles Boyd has one quarter-sized piece left of Einstein's brain, the other having been sacrificed to the blender. He grew uncomfortable after the failed effort to run a paternity test, sensing that perhaps he had inadvertently landed in the midst of a family feud. He has no plans to do anything with the final piece. People sometimes ask if scientists might one day be able to salvage enough DNA to clone Einstein's brain. The query makes Boyd think of the woolly mammoth discovered in the Siberian tundra in 1997 that his colleagues are trying to resurrect, filling in the missing gaps of DNA with elephant genes to implant it in the womb of a surrogate elephant mother. "There's something so much like science fiction about all of this," he said. When he learned that Einstein's eyes remain preserved in a New Jersey safety deposit box, he speculated that the optical tissue might indeed be a better source for extracting DNA. "It's an intriguing possibility and our interest in doing this would depend very much on why one would want to do this. The only real reason I can think of would be to revisit Evelyn's paternity; this is such an explosive issue that I think it would be best handled by a forensic lab." But, he said, it does raise the question as to whether scientists could find within it a gene related to high intelligence (as British scientists first reported they had done in 1997).

For his own research purposes, the Einstein project turned out to be a terrific success for Boyd, nonetheless. When *The Economist* ran a brief item about his efforts to find a gene associated with aortic aneurysms in Einstein's DNA, a woman in Louisiana read it. "She came from a large family with a genetic mutation and a strong history of aortic aneurysms. It was hugely helpful," Boyd said. "She actually went on to establish an outreach program for other families." In 1995, he and his wife moved to the University of Hawaii to continue studies of collagen- and elastin-related disorders and they took the vial containing the remaining nub of Einstein's brain with them. "The tissue is soaking in ethanol in the fridge. I peeled off that sticker that said Big Al's brain," he said. "I kept it because it would be a travesty to throw it out. I hardly show it to anyone, though I did once to a neuro-surgeon and the occasional student, and they all get glassy-eyed. Harvey preserved it for science—he was doing it for the most remarkable brain ever produced by this world and we have one small part of it and we are in awe of it." Since meeting him that one day in Princeton, Boyd has always felt a little sorry for the old pathologist. "He seemed weighed down. Obviously it was a great idea to take the brain of the planet's smartest man, but then he was caught in a subtle balance between preserving it for science or reverence: some-thing to gawk at versus something to study."

✶

Harvey inspected the jars one day in 1996 and decided it was time; the alcohol level had dipped dangerously low. He climbed the steps from the basement to the phone in the kitchen. A few days later, he pulled on his coat and heaved the box upstairs, pushed open the door and trudged across the deck, then down the steps to the car. He rolled down the long stretch of his Titusville driveway, past the little mail-box with the names Wheatley and Harvey stuck to the side, and turned toward town.

Retired doctors often drop by the office of Elliot Krauss. He has a good sense of humour. They tell dirty jokes and racy stories about the old days at Princeton Hospital, the naughty affairs with nurses and assorted Peyton Place tales of the 1950s. Someone, at some point, he was sure, had mentioned the strange things that happened up the hall the day that Einstein died. But not until that afternoon, when the phone rang, did he realize the story was true.

"Dr. Krauss?"

"Speaking."

"My name is Thomas Harvey and I used to do your job about fifty years ago. I was wondering if I could come in to get some alcohol, you know, so it doesn't dry out."

Krauss hid his surprise. "Sure, sure, come on in," he said.

Just as Jack Kauffman had always imagined it could be, the hospital had grown into a full medical centre. Extensions crowded out the asphalt that once afforded doctors ample space to park. Now most visitors had to find a spot on the street. With the box in his hands, it must have been a strain for the eighty-four-year-old Harvey to reach the entrance. The front steps where he'd stood a lifetime ago talking to reporters had been replaced by sliding doors that open into a spacious lobby. To the left was a raised platform of tables and chairs, which never existed in his day, where patients and their families ate off trays and drank coffee. Below it was a bank of pay phones, and beside that a counter where white-haired ladies in pink coats gave people directions. Harvey didn't need any. He navigated his way to a door on the right, somehow managed to push it open and smiled in a kindly way at the nurses, who told him to go straight in. "He had a certain charm all right," Krauss would say. "The nurses found him endearing. They mentioned he had these nice blue eyes."

Krauss sat behind the desk, a tall, trim forty-year-old with a thick moustache and the remnants of a golf course tan, surrounded by bookshelves stuffed full in a windowless office with two doors. One of them opened to the polished-concrete corridor that led to the green-tiled room where it all began. The other door opened to the nurses' counter, where Harvey appeared that day, his thin arms straining around the girth of a cardboard box from the Kaweah Citrus Association. "He looked like a wizened little old guy, like anyone's grandfather," Krauss thought. Then he read the numbers scrawled in green magic marker on the box: "'55-33." The younger pathologist knew immediately what they meant: 55 for the year and 33 "for the number of the autopsy report, the thirty-third autopsy done that year. I went to look for the report once, but it had disappeared."

From the box Harvey pulled up the litre-sized glass jars. Tracks of sticky white residue smeared the lids from the years of peeling back the masking tape to retrieve the pieces Harvey had taken out to admire or plop in a mayonnaise jar, in a courier kit, in a Ziploc bag, into the anxious hands of a Japanese math professor and so many other places besides. Krauss could see the pieces protruding above the dwindling yellowish liquid. Some were wrapped in gauze, and all were suspended within their waxy, golden prisons. "As soon as I saw it I knew this preservation was dramatically different than others I'd seen," Krauss said. "Nobody did what he did to preserve a brain. Usually you take it and dump it in formalin for two weeks until it sets and then you can slice it right in half. I figured he knew he was going to keep it for posterity and had it magnificently preserved."

Krauss complimented Harvey on the preservation of his specimen. Harvey took a shine to the young pathologist. He sat down in the chair opposite the desk and shared some of the details of the preservation process, describing how a German woman in Philadelphia had so deftly embedded its 240 pieces. Krauss knew there had been controversy over Harvey's taking Einstein's brain, but later he said, "When you give permission for an autopsy you are giving permission for a pathologist to study all the organs as well. The philosophy, especially at teaching centres, is that if you see an organ and it's interesting, you keep it for learning purposes." Even as the two of them sat there on that first afternoon, the old morgue—one of the few unrenovated rooms in the institution—was stocked like a pantry with the pickled body parts of former patients: hearts, livers, and even a few slices of brain, unclaimed and unlabelled on the old steel shelving unit.

Later, Harvey could not recall when the thought had first occurred to him. It might have been that the seeds were planted that first day

when he asked Krauss how it was that he had come to be the chief
pathologist at Princeton Hospital. Unlike Harvey, Krauss had not
come to the profession by illness or accident. He'd chosen it, had an
inclination since boyhood. His father toiled in a lumberyard. His
mother worked as a secretary for a local doctor who used to feed him
back issues of *Scientific American* when he was a boy. The doctor also
introduced him to anatomy with multicoloured plastic models of
human organs that could be pried apart and put back together like
Lego. Krauss assembled and reassembled them in his childhood home
minutes down the pike from Manhattan in Teaneck, New Jersey, a
post-war suburb of perfectly angled yards and lawns mowed too short
for weeds to go unnoticed. Harvey's ears perked up when Krauss
told him he'd studied a bit of neurophysiology at the University of
Rochester, that he got a charge out of implanting electrodes in rat
brains. Probably, too, the fact that he admitted that pathology was
the only course he'd finished with honours in his second year at medical school appealed to Harvey's Quaker sensibilities: Krauss was a
modest man. Other doctors assumed that you had to be anti-social
or a complete oddball to want to spend your career peering down a
microscope at people's cells, or cutting open corpses when it's too
late to fix the problem you find inside, Krauss said, but he'd had fine
professors who told him otherwise: John Kissane and Lauren
Ackerman and Juan Rosai, pioneers of surgical pathology in America.
He'd tried a year as a hospital resident, but he was a newlywed then,
and he'd wanted nights home with Mary. It was funny that they'd
ended up in New Jersey, Krauss said. Mary had ranked it fifty-third
on the list of states in which she wanted to live. But they'd been driving through, on their way to a friend's wedding in Philadelphia, and
Krauss was hunting for his first job, so he'd stopped for an interview
at the Robert Wood Johnson Medical Center in New Brunswick,
twenty-five minutes by car from Princeton. They'd stayed there for
seven years. At thirty-eight, Krauss said, he was a long shot for the

Princeton chief pathology job. But a plum post like that "never really comes up," so he applied. He and Mary and their two boys moved to Princeton in 1989 and never looked back. "I'm a bit of a history buff," he said, "and this old town is a bona fide cradle of America."

Harvey became one of the regular visitors to the pathology department. He dropped in on Krauss if he had to see his doctor, or sometimes just to chat if he happened to be in town. Two years passed after their first meeting before the subject came up. Harvey said he had been thinking about it—"Krauss is a nice man and he told me he had done some early work in neurology"—but it was Krauss who phoned. He didn't tell the hospital officials he was going to do it. He didn't even tell Mary. He worried that it might sound presumptuous, but he called anyway.

"You are not getting any younger," he told Harvey, "and I'd like you to give some thought to giving the brain back to the Princeton Medical Center . . ."

"Okay," Harvey said, "I'll think about it."

"I personally think Harvey did the right thing. It would not have been preserved without him. We have to believe science will ultimately find something useful in it because I have faith in science, and we have to because it is what we do." Why did Harvey want to saddle himself with the brain of Albert Einstein in the first place? "I think everyone wants eight seconds of fame," Krauss said, recalibrating Andy Warhol's famous adage to the sound bite era. "I think people go about their business and live most of their lives in anonymity, but then a chance comes along for nearly everyone to have that eight seconds of fame, and they want it, maybe as something to leave behind in their name, maybe to know their life was worth something . . . I think that's what it was for him . . . I guess in a way, that's what it was for me."

By Harvey's clock, he didn't think long. Even there, in the rural hills of Titusville, a scientist from Calcutta had found him and sent him a letter asking for a piece of it. And he'd recently shipped off a piece to a neurologist in South America who wanted to study Einstein's glial cells. "I really wanted to get out from under all that responsibility. I had supervised its disposition for many, many years," Harvey said. "Anyone who had Einstein's brain to take care of would have taken good care of it. But you see, I had made a promise and I did try to take it very seriously, I hope I did that.

"I didn't know how long it would take at that time to do studies, but I have no regrets about keeping it this long, you know, before it is completely studied. There is still lots to be done." But now, Harvey decided, it was someone else's turn. And so of all the possible big-name brain researchers and prestigious institutions, it would be Elliot Krauss, who, by some alignment of distant stars, happened to be the one in the right place at the right time—or, as Krauss would later joke, "the wrong place at the wrong time, depending on how things turn out." It wasn't so much that it was Krauss specifically, Harvey said, though of course he was fond of him. "It was where he was," Harvey explained. "I guess that's where the brain started out, where the autopsy was, where I had been, working."

Two weeks after Krauss had telephoned to ask for it, Harvey packed up his homemade maps with the numbered pieces and carried Einstein's brain out to the car for the last time. "I felt like I was nearing the end of a very long trip," he said. Who knows what thoughts skipped through his mind as Harvey drove the thirty kilometres from Titusville to Princeton, whether he had doubts or drilled the ball of his foot hard against the accelerator to have it, at long last, be over, to be free?

Once again, Harvey appeared in the doorway. Krauss was amazed to see him. He would be near giddy and half frantic in the days to come, wondering where he should lock up the brain, whom he should tell. "I worry there might be some nutcase out there and I worry about a break-in, or fear for my wife and family and generally try to keep a low profile worrying someone might try to steal it." Still, he would be flush with the promise of it, discussing the miracles of technology, holding out the possibility of harvesting Einstein's DNA in the future—maybe even cloning it. But he wasn't thinking that then; not at the moment Harvey stood before him hugging the box, ready to pass his life's prized possession and mission into *his* hands.

If the old pathologist felt at all sentimental, he never said. Harvey simply set the box down on the floor and disappeared.

EPILOGUE

I SAT ACROSS the desk from Elliot Krauss on a frosty February morning in 2000. We'd been speaking in his boxy office at the Princeton Medical Center for more than an hour when he asked me.

"Do you want to see it?"

"You have it here?"

He grinned and nodded.

He wouldn't say exactly where. It wasn't the kind of information a person wants to spread around, he explained.

This wasn't the first time Krauss had shown it to a visitor. The media parade had begun two days after Thomas Harvey made him heir to Einstein's brain. In the two years since then, envoys from the *Princeton Packet*, the History Channel and the BBC had come calling. Krauss had the drill down pat. He requested that I step out into the hall for a few moments while he fetched it from its secret location.

I waited in the main floor corridor, opposite the senior citizen volunteers at the information desk, wondering how many times Krauss would tell the story before his shoulders hunched and the years shaved centimetres off his height. Friends and relatives and strangers would no doubt keep asking. And as sure as editors dispatch reporters to cover summer carnivals and spiking crime statistics year after year, someone from somewhere, even decades from now, would be quizzing the good-humoured Princeton pathologist to find out whatever became of it.

If Krauss has his way, the prized specimen won't look terribly different in the future from the day he inherited it. As he told me in his office, reclining casually with his hands clasped behind his head, he has a plan. He intends to treat the chopped organ like books in a library: scientists can check pieces out to study and return them when their research is complete. So far, Krauss said, the arrangement has worked well with a group in Japan examining Einstein's brain for signs of Alzheimer's disease and a neurobiologist in Buenos Aires studying the late physicist's glial cells in half a dozen blocks from Broca's area.

Harvey gave Krauss his homemade maps of the organ. But the checkmark-and-shading system seemed to Krauss outdated. As with any change in management, the new custodian has developed his own inventory forms to keep track of the cerebral chunks coming in and going out. "You know," he said, "I didn't realize it would be such an enormous responsibility." Yet, like his predecessor, Krauss doesn't accede to all requests. In April 1999, officials at the Deutsches Hygiene-Museum in Dresden wrote to him asking if they could display Einstein's brain for their exhibition on multiple intelligences. Krauss turned them down. Although it has appeared alongside him on camera, the brain, he said, is to be the object of scientific study only. And even in those matters he exercises discretion. An assistant from Sandra Witelson's office has called a few times requesting more tissue, he said. But Krauss was unimpressed with the paper in *The Lancet*. He believes that if there are differences to be found in Einstein's brain they will be discovered at the cellular level —not in the shape or size of the organ. "I think that sort of stuff is glorified phrenology," he said.

All indications are that Witelson has shifted her Einstein work from measurements to microscopes. Although she decided not to be interviewed for this book, as Witelson said she hopes one day to tell her own story, she has in other venues alluded to her research plans. At the October 1999 Society of Neuroscience meeting in Miami, Witelson reported that she'd found that the more densely cells are packed into a particular region of the brain, the more intelligent the person. As she told me when I interviewed her for a *Globe and Mail* story at the time: "The closer the cells were together in a bit of tissue—like granules in a cube of sugar—indicated how well the subjects performed [in the intelligence tests]." At the 1997 neuroscience meeting in New Orleans, Witelson, along with her assistant Debra Kigar and New York neuropathologist Ilya Glezer, reported that they had assessed and counted Einstein's brain cells in the slides from regions believed to be involved in language, which are usually found around the ears, just below the Sylvian fissure. Given that Einstein was thought to have "unique features in speech production," they used a computer-assisted microscope to compare Einstein's cells in these regions to the same brain areas in eight men of normal intelligence. But according to the abstract they submitted, they found nothing remarkable in terms of numbers of neurons or their density. One outstanding question is how dense are the cells in Einstein's brain in the areas thought to be involved in his particular talents.

But Krauss and Witelson appear to have gotten off on the wrong foot. When I asked him if he had ever spoken to the Hamilton neuropsychologist, he shot back: "*She* has never spoken *to me* . . . I've heard from her assistants." Krauss said he doesn't plan to send Witelson any more tissue: "She's got enough."

For Witelson, and any other scientist denied access to the Einstein specimen, there is no one to whom they can appeal the decision. There is no academic research committee, or higher authority or even an Otto Nathan to hear their case. Once again, a single, small-town pathologist is left to decide the fate of history's most celebrated brain. Not even Harvey keeps tabs.

In nearly forty-five years, Harvey never did tell anyone about his relationship with Otto Nathan. Harvey's claim that Nathan had watched his handiwork in the morgue that April morning in 1955 was the only suggestion that the pathologist had not acted alone in keeping Einstein's brain. Instead, Harvey—who has earned not a cent from his spectacular specimen—bore the stigma of a thief.

Yet all the while he had in his possession, along with the brain, at least a few of Nathan's letters to prove that he was acting with the knowledge and consent of Einstein's estate. These letters, which were among a hand-selected pile that Harvey offered me from his mountainous collection, marked the beginning of a trail that led me to the Einstein archives, and eventually to the executor's stash, which shed new light on the story. To this day, I cannot be certain if Harvey included them deliberately. But he seemed not at all surprised when I told him what I'd found. "Oh," he said, "maybe they will help refresh my memory."

His family meanwhile feel compelled to speak up in his defence. As his son Arthur Harvey told me, "He kept the brain with him all these years and he came under criticism for it and its preservation. He wanted, I think, to do the study himself, but he never really had the time. Life kind of intervened, and he was busy raising two families and supporting them. I think it was a dream that was never realized. But it is well preserved and he is responsible for that."

With Paramount Pictures planning to make a movie of Michael Paterniti's road-trip book, friends and family worry about Harvey's portrayal. "They think it might not be too complimentary to me," Harvey said. Despite his limp and his eighty-eight years, he cut a mean pace as I walked by his side in downtown Princeton last summer to the office of the man who has been his lawyer for more than thirty years—the man who is now handling his negotiations with Hollywood. This, as with everything else, has landed in Harvey's unlikely lap. And as always, he seems not the least ruffled or affected by it.

Harvey fills his days with woodworking and long drives with Cleora to Virginia to visit historical sites. Sometimes he'll head all the way to Baltimore if he hears that Ahilleas Maurellis—who now lives in the Netherlands—happens to be in town for a conference. Every

Wednesday, he climbs on the Princeton train to Philadelphia to volunteer his services at the Quaker Information Center, sitting by the window, whizzing once again by the countryside. Then he rides it home again to Titusville, to the front parlour with its eleven reading lamps and its Queen Anne end tables, his piles of magazines and books straining their dainty legs. There's a copy of *Der Spiegel*, the German magazine that featured an article on Einstein's brain two years back, and a hardcover edition of John D. Rockefeller's *Random Reminiscences of Men and Events*. Harvey says he still plans to write his own book—an atlas of Einstein's brain—now that he has the time up on his hillside retreat, invisible from the winding road below.

I had lingered too long in the hallway and Krauss eventually poked his head out and motioned that it was time for me to come in. "It's kind of anticlimactic, isn't it?" Krauss said as he towered over it. He'd cleared space on his desk so I could see what all the fuss was about. And there it was, the immortal brain of Albert Einstein that once pulsed, fired and changed the world, now in two cookie jars that looked like fish tanks in need of cleaning. Wrapped in muslin cloth, numbered stickers still attached, damp and unglorious wedges still in the dubious service of science.

SOURCES

EPIGRAPH

Einstein's comment after his first trip to the United States is quoted with permission from the Albert Einstein Archives, the Jewish National and University Library, the Hebrew University of Jerusalem. It is taken from "Einiges ueuber meine Eindruecke in Amerika," in *Mein Weißbild*, 1934.

CHAPTER ONE

Details of the discussion between the late Harry Zimmerman and the medical resident at Montefiore Hospital were relayed to me in interviews in January and September 2000 with Roberto Salvatori, now an endocrinologist at Johns Hopkins University.

In rounding out the biographical details of Harry Zimmerman's life and career, I relied on various obituaries, in particular Asao Hirano and Leopold G. Koss, "Harry M. Zimmerman 1901–1995," *Acta Neuropathology* (1995) 90:545–46; Robert McG. Thomas Jr., "Dr. Harry Zimmerman, 93, Dies; Founded Albert Einstein College," *New York Times*, July 31, 1995, p. 7. Interviews with neuropathologist Asao Hirano at the Albert Einstein School of Medicine, Robert Terry, emeritus professor of neuroscience and pathology at the University of California, San Diego, and Bertram Lincoln Pear, clinical professor of radiology at the University of Colorado, also supplied information. Dr. Pear gave permission to quote from a letter Harry Zimmerman wrote to him on June 24, 1975, regarding Einstein's "genius cells."

To re-create the conversation Zimmerman had with Albert Einstein in naming the new medical school, I drew on details Zimmerman himself included in an article that he wrote entitled "The Naming of the Albert Einstein College of Medicine," *Surgical Neurology* (August 1976), vol. 6, no. 2, and Ernst. R. Jaffe, "The Early History of the Albert Einstein College of Medicine," *The Einstein Quarterly Journal of Biology and Medicine* (1996), vol. 13. An interview in February 2000 with retired physician Thomas Bucky, who was present at the 1953 meeting, was also valuable.

To depict Harry Zimmerman's views concerning the continuing study of Einstein's brain, I relied on various newspaper and magazine accounts in which Zimmerman has been quoted over the years, among them: Chris Szechenyl, "Einstein's brain still a convoluted puzzle," *Kansas City Times*, October 27, 1981, p. 1; Gina Maranto, "The Bizarre Fate of Einstein's Brain," *Discover*, May 1985, p. 28; and "Can Scientists Link Einstein's Genius to his Brain Shape?" *Detroit News*, August 8, 1993, p. 1.

Since the fetal development of Albert Einstein's brain can never actually be known, I used various sources to compile a general description of the embryonic formation of the organ. These sources include Arnold B. Scheibel's 1998 article "Embryological Development of the Human Brain," posted on the online *Brain Lab*, maintained by New Horizons for Learning (copyright 1997 through 2001); J. Madeleine Nash, "Fertile Minds," *Time*, February 3, 1997; Lewis Wolpert, *The Triumph of the Embryo* (Oxford: Oxford University Press, 1991); Peter W. Nathanielsz, *Life Before Birth and a Time to Be Born* (Ithaca, NY: Promethean Press, 1992); and Christopher Vaughan, *How Life Begins: The Science of Life in the Womb* (New York: Times Books, Random House, 1996). Information about the state of brain cells after death came from Theo D. Palmer et al., "Progenitor Cells from Human Brain after Death," *Nature* (May 2001), vol. 411.

For information on Einstein's life and science I owe a great debt to biographies of the Nobel laureate, in particular: Denis Brian, *Einstein: A Life* (New York: John Wiley & Sons Inc., 1996); Ronald W. Clark, *Einstein: The Life and Times* (New York: Avon Books, 1972); Abraham Pais, *Einstein Lived Here* (New York: Oxford University Press, 1994); Banesh Hoffman with Helen Dukas, *Albert Einstein: Creator and Rebel* (New York: Viking Press, 1972); and Roger Highfield and Paul Carter, *The Private Lives of Albert Einstein* (London: Faber and Faber, 1993). The reference to Einstein's EEG originated with the Associated Press and appeared under the headline "Geniuses Aid Tests of Brain Processes," *New York Times*, February 24, 1951, p. 9; Stephanie Sammartino McPherson, *Ordinary Genius: The Story of Albert Einstein* (Carolrhoda Books, 1997). Interviews in February and June 2000 with

Charles Dyer, professor of astrophysics at the University of Toronto, also helped me to lay out some of Einstein's scientific concepts.

CHAPTER TWO

The anecdote from Valley Road School came from Katrina R. Mason, an author and teacher now living in Maryland. Ms. Mason, whom I interviewed in January 2000, also supplied details about life in Princeton in the 1950s, in particular a child's view of Einstein.

Information about the Harvey family history and Harvey's education, personal and professional experience came from various interviews with Thomas Harvey, now living in New Jersey, over the two-year period from June 1999 to June 2001. References to the history and philosophy of Quakerism came from various sources, among them: discussions with Thomas Harvey; the many educational essays posted online under *The Religious Society of Friends* (at www.Quaker.org) and *The History of Quakerism*; Laurence Barber, "Know the Quakers" (copyright 1990, 1999), posted online by the *Religious Society of Friends, Quakers on Cape Cod.*

Obituaries previously listed and interviews with Asao Hirano and Robert Terry provided background on the late Harry Zimmerman's career.

To assemble the information about the events and personalities that drove early brain science, Stanley Finger, *The Origins of Neuroscience: A History of Explorations into Brain Function* (New York: Oxford University Press, 1994) was a valuable tool, as was information Finger shared with me during an interview in December 2000. As well, the timeline included in "Milestones in Neuroscience Research," posted by Eric H. Chudler of the Department of Anesthesiology at the University of Washington, and *The History of Phrenology on the Web* (copyright 1999 through June 2001), by John van Wyhe of the Faculty of History at the University of Cambridge, were particularly helpful.

Details about Oskar and Cecile Vogt and the analysis of Lenin's brain came from various sources: Andrew Higgins, "Vladimir's Brain," *The Age*, November 20, 1993, p. 5; M. Bentivoglio, "Cortical structure and mental skills: Oskar Vogt and the legacy of Lenin's brain," *Brain Research Bulletin* (November 1998), vol. 47, no. 4; Walter Reich, "Pickling the Brains of Geniuses," *San Francisco Chronicle*, August 11, 1985, p. 13.

To round out details of Harvey's stay at the Gaylord Sanatorium, I relied on *The Spirit of Gaylord: A History*, compiled by medical historians and volunteers (New Haven, CT: Cheney & Company, 1997).

Harvey Rothberg, *The First 50 Years: A History of Princeton Hospital, 1919–1969* (Princeton, NJ: Medical Center at Princeton Foundation, 1995) provided valuable insight about Princeton Hospital's beginnings and the people who shaped it. Details that brought the history to life came from interviews conducted from January to April 2000 with doctors who worked at the hospital during this period, including Louis Fishman, D. Barton Stevens, Robert Lewis, Benjamin Wright, David Rose and Henry Abrams. Dr. Abrams also recounted for me, in February 2000, memories of his friendship with Albert Einstein.

Past issues of the *Princeton Packet* and Richard D. Smith, *Images of America: Princeton* (Charleston: Arcadia Publishing, 1997) also offered valuable historical colour and facts.

Interviews with Arthur Harvey and Robert Harvey in Princeton in February and March 2000 supplied some of the personal details of their family life there.

CHAPTER THREE

The anecdote involving the New York artist and photographer from Hearst Metrotone News came from the article entitled "Dr. Einstein's Own Camera Used to Snap Only Photo on Birthday," *Princeton Packet*, March 17, 1955, p. 1.

To reconstruct details of Einstein's final days I relied on passages from: Denis Brian's *Einstein: A Life*; Ronald Clark's *Einstein: The Life and Times*; Roger Highfield and Paul Carter's *The Private Lives of Albert Einstein*; Jon R. Cohen and L. Michael Graver, "The Ruptured Abdominal Aortic Aneurysm of Albert Einstein," *Surgery, Gynecology & Obstetrics* (May 1990), vol. 170; a letter by James J. Chandler that included unpublished notes from the files of Dr. Guy Dean, published in "The Einstein Sign: The clinical picture of Acute Cholecystitis by Ruptured Abdominal Aortic Aneurysm," *New England Journal of Medicine*, June 7, 1984, p. 1538; "Dr. Albert Einstein Dies in Sleep at 76; World Mourns Loss of Great Scientist," *New York Times*, April 19, 1955, p. 1; interviews with retired physician

Thomas Bucky, who visited Einstein in the hospital the weekend before he died, and former hospital surgeon D. Barton Stevens.

Biographical material about Otto Nathan came from interviews through the summer of 2000 and into the spring of 2001 with: Thomas Bucky; Robert Schulmann, professor of German history and an editor of *The Collected Papers of Albert Einstein*, now based at the California Institute of Technology; and Jamie Sayen, former neighbour to Albert Einstein who came to know the late executor while writing *Einstein in America: The Scientist's Conscience in an Age of Hitler and Hiroshima* (New York: Crown Publishers, 1985). Information about Nathan's battles with the U.S. State Department came from various articles that ran in the *New York Times* and an online excerpt from Otto Nathan in the article by Ruth and Bud Schultz, "Wasn't that a time? A Century of Struggle. A Century of Repression," posted on the *Human and Constitutional Rights Resource Page* at Columbia University Law School.

Information about the autopsy suite at Princeton Hospital came from a tour of the room in February 2000 and an interview with Elliot Krauss, now chief pathologist at Princeton Medical Center. To reconstruct events of the autopsy itself, I relied on interviews with Thomas Harvey, details of the Virchow method as explained in G. A. Gresham and A. F. Turner, *Post Mortem Procedures* (London: Wolfe Medical Publications, 1979) and witnessing a variation of the Virchow method at the Centre for Forensic Sciences in Toronto in August 2000.

Information about Henry Abrams's role and the scene inside the autopsy suite the day Einstein died came from an interview I had with Abrams in February 2000. For the passage involving relics, I drew on material and inspiration from various sources, among them: the Catholic Online website, a member of the Catholic Press Association (copyright 1990–2001); E. Richard Gold, *Body Parts: Property Rights and the Ownership of Human Biological Materials* (Washington: Georgetown University Press, 1996); Jonathan Freedland, "In the name of Science," *The Guardian*, December 17, 1994, p. T010; Nino Lo Bello, "Search for Galileo's Finger Leads Way to Sights Far Beyond Package Tour," *Chicago Tribune*, January 26, 1986, p. 3.

CHAPTER FOUR

The reconstruction of events in the days following Einstein's death were based on interviews with Arthur Harvey (February 2000), Louise Sayen, Jamie Sayen and Thomas Bucky (May 2001), Robert Terry (March 2000), Robert Schulmann and Thomas Harvey. Information that appeared in print at the time came from: "Dr. Einstein Dies in Sleep at 76 . . ." *New York Times*, April 19, 1955, p. 1; "Dr. Einstein's Will with Bequests for $75,000 Filed in Freehold," *Princeton Packet*, May 1955; "Son Asked Study of Einstein's Brain," *New York Times*, April 20, 1955, p. 24; "Key Clue Sought in Einstein Brain," *New York Times*, April 20, 1955; "Einstein Study Called," *New York Times*, April 22, 1955; "Ottumwans' Nephew to Direct Einstein Study," *Ottumwa Daily Courier*, April 21, 1955; "Doctors Row over Einstein's Brain," reported by the Associated Press, *Chicago Daily Tribune*, April 20, 1955; "Brains of Great Scientist Cause Hospital Dispute," reported by the Associated Press, April 21, 1955. The office staff of the Ewing Crematorium supplied information about Nathan paying the bills in May 2000.

Sources and inspiration for the section about how pathologists treat corpses and the blurred lines of informed consent in the autopsy suite came from Ruth Richardson, *Death, Dissection and the Destitute* (New York: Penguin Books, 1988) and Lori Andrews and Dorothy Nelkin, *Body Bazaar: The Market for Human Tissue in the Biotechnology Age* (New York: Crown Publishers, 2001).

For information related to the strained relationship between Albert Einstein and Hans Albert Einstein, Roger Highfield and Paul Carter's *The Private Lives of Albert Einstein* was invaluable, as was information provided me by Evelyn Einstein during interviews in March 2000 and January 2001.

The chapter on intellect and great brains in Stanley Finger's *Origins of Neuroscience* was a wonderful resource in chronicling the early attempts to correlate brain biology to intelligence. Stephen Jay Gould, *The Mismeasure of Man* (New York: W. W. Norton & Company, 1981) and Nigel Hawkes, "The thinking man's brain teaser," *The Times* (London), June 19, 1993, also supplied details. Information about Wilder Penfield's work came from "Wilder Penfield 1891–1976," posted on the People and

Discoveries site of *A Science Odyssey*, a link from PBS online (copyright 1995 through 2001), and the review by Rick Groen entitled "Engrossing NFB documentary on a brilliant man: Penfield's life-long quest," *Globe and Mail*, Nov. 21, 1981, p. E1.

Letters written by Otto Nathan are quoted with the permission of the Albert Einstein Archives at the Jewish National & University Library, The Hebrew University of Jerusalem, Israel.

Information about the brain of Benito Mussolini is based on Rod Moran's article "Mussolini Under the Microscope," *West Australian*, June 19, 2000, p. 3. Biographical details about the late Webb Haymaker were kindly supplied in a mailing by Louis D. Boshes, professor of neurology emeritus at the University of Illinois at Chicago, and they include a reprint of his article "Webb Haymaker: Operation Stratomouse," *Journal of the History of the Neurosciences* and Francis Schiller's "Webb Edward Haymaker, 1902–1984," *Neurology* (March 1985), vol. 35.

The brain maps that Thomas Harvey sketched and Percival Bailey and Gerhardt von Bonin's *The Isocortex of Man* (Urbana, IL: University of Illinois Press, 1951) helped to explain the method Harvey followed to section the specimen. The exact measurements he recorded were taken from the numbers reprinted in Sandra Witelson, Debra Kigar and Thomas Harvey, "The Exceptional Brain of Albert Einstein," *The Lancet*, June 19, 1999.

CHAPTER FIVE

The bulk of this chapter is based on interviews with Thomas Harvey.

Information about celloidin and the embedding process came from various sources: the helpful responses I received online through *The History of Neuroscience Internet Forum*, maintained by the History of Neuroscience and Brain Research Institute at the University of California at Los Angeles; interviews with Robert Terry and Nicholas Gonatas, professor of pathology and laboratory medicine at the University of Pennsylvania, in April 2000; the assistance of Wally Welker, professor of physiology and neuropathology and a founding director of the Comparative Mammalian Brain Collection at the University of Wisconsin; and the critical eye of Inge Sigglekow, senior histologist and manager at Wisconsin's department of neurophysiology.

The correspondence between Thomas Harvey and Otto Nathan is quoted with permission from the Albert Einstein Archives in Jerusalem.

Sidney Schulman, professor emeritus of medicine and neuroscience at the University of Chicago, shared details of his encounter with Einstein's brain and his dealings with Harvey during an interview in April 2000.

In writing the section about the reluctance to study intelligence in the 1950s, I drew on information I gathered in interviews with: Sidney Schulman; Albert Galaburda, professor of neurology and neuroscience at Harvard Medical School and chief of behavioural sciences at the Beth Israel Deaconess Medical Center, in February 2000; and Marian Diamond, professor of neuroanatomy at the University of California at Berkeley, in March 2000. I also found inspiration in Tom Wolfe's article "Sorry, but Your Soul Just Died," *Forbes*, December 2, 1996.

To describe Percival Bailey's look at the slides from Einstein's brain, I relied on information Louis D. Boshes provided in an interview in February 2000.

Details about the brain at the Harvey house in Princeton came from interviews with Arthur Harvey and Robert Harvey.

Stuart Fischoff, media psychologist in the School of Natural and Social Sciences at the California State University at Los Angeles, provided historical perspective on the media's reaction to the story of Einstein's brain in a June 2001 interview.

CHAPTER SIX
Interviews conducted with Thomas Harvey, Robert Harvey, Robert Schulmann, Evelyn Einstein, along with correspondence from the Albert Einstein Archives and Roger Highfield and Paul Carter's *The Private Lives of Albert Einstein*, provided information to assemble this chapter.

Background on the late Kurt Goldstein came from "Kurt Goldstein (A Biographical Note)" published in the bi-annual newsletter *In Context* (The Nature Institute, Fall 1999).

Details about Harvey's departure from Princeton Hospital and information about the upheaval at the hospital itself were supplied, during interviews held through February and April 2000, by several retired doctors who worked at the institution during those years, among them Louis

Fishman, Robert Lewis and David Rose. Harvey Rothberg's book, *The First 50 Years: A History of Princeton Hospital, 1919–1969*, also provided specific dates and details of certain exchanges that took place.

Hospital staff records concerning Thomas Harvey's departure came from the Princeton Medical Center, and Harvey gave me permission to quote from these files.

The last line of the chapter regarding Elouise Harvey's statement to co-workers came from Denis Brian's *Einstein: A Life*.

CHAPTER SEVEN

To compile the setting and scene for this chapter, I relied on information gathered on a trip to the University of California at Berkeley in March 2000, which included an interview with Marian Diamond. In order to fit the text intended for the general public, this presentation of her work— which spanned decades—has been condensed to give an overall sense of her findings. Information about Arnold B. Scheibel, professor of neurobiology at the University of California at Los Angeles, came from my interview with Scheibel in February 2000.

Information on the article that graduate students tacked to Diamond's lab wall came from "Brain That Rocked Physics Rests in Cider Box," *Science*, August 25, 1978, p. 696.

Bruce Ransom, professor of neurology at Washington University School of Medicine, provided helpful background information in January 2001 on the unglamorous history of glial cells, as did the article by Jamie Talan, "Move Over Neurons . . . Out of the dark glial cells take center stage in brain research," *Newsday*, October 10, 2000, p. C03.

Anecdotes about the problems that can afflict people with parietal lobe damage came from Jay Ingram, *The Burning House: Unlocking the Mysteries of the Brain* (Toronto: Penguin Books, 1995); V.S. Ramachandran and Sandra Blakeslee, *Phantoms in the Brain: Probing the Mysteries of the Human Mind* (New York: William Morrow, 1999); and Frederick E. Lepore, "Dissecting Genius: Einstein's Brain and the Search for the Neural Basis of Intellect," *Cerebrum* (Winter 2001), vol. 3, no. 1.

Information about Korbinian Brodmann came from the online postings at the *Founders of Neurology* from the Louis D. Boshes, M.D., Archives at the University of Illinois at Chicago, and from the essay posted online about Brodmann's work by Laurence J. Garey, professor of anatomy at the Imperial College School of Medicine in London, who also translated from German into English Brodmann's *Localisation in the Cerebral Cortex* (London: Smith-Gordon, 1994).

CHAPTER EIGHT

Information about Harvey's childhood summers in the Midwest came from interviews conducted with Thomas Harvey. Quotes related to Harvey's conversation with reporter Steven Levy were taken from Steven Levy, "My Search for Einstein's Brain," *The New Jersey Monthly*, August 1978, p. 43.

Information about developments in brain science during the 1970s came from various sources, among them: an interview with Dr. Bert Pear of Denver, Colorado; Jay Ingram's *The Burning House*; Bruce Bower, "Whole-brain interpreter: a cognitive neuroscientist seeks to make theoretical headway among splitbrains," *Science News*, February 24, 1996, p. 124; Joan Hollobon, "Discoverer of 'happy' chemicals wins award," *Globe and Mail*, November 2, 1978, p. T2; Joan Hollobon, "Everyone knows it hurts and it's costly but the things that trigger torment are still largely unprobed mysteries," *Globe and Mail*, p. P10.

Information about Michael Aron's correspondence with Otto Nathan came from the Albert Einstein Archives in Jerusalem, as did all letters between Nathan and Harvey from this period.

To chronicle the efforts of Nathan and Dukas in preserving Einstein's legacy I relied on interviews with Robert Schulmann and Jamie Sayen, and on Roger Highfield and Paul Carter's *The Private Lives of Albert Einstein* and Denis Brian's *Einstein: A Life*.

CHAPTER NINE

Events related to Einstein's trips to California came from Denis Brian's *Einstein: A Life* and Ronald Clark's *Einstein: The Life and Times*. Interviews with Charles Dyer, professor of astrophysics at the University of Toronto and an expert on relativity, also helped to put into context Einstein's handling of the cosmological constant.

Interviews with Arnold Scheibel provided information about his hopes to study Einstein's brain.

Information about Weston, Missouri, came from Joel M. Vance, "Sin city: Tiny Weston, Mo., founded on alcohol and tobacco staggers back to life," *Chicago Tribune*, March 3, 1991, p. 16. Information about Harvey in Weston came from: Chris Szechenyl, "Einstein's Brain Still a Convoluted Puzzle," *Kansas City Times*, October 27, 1981, p. 1; Nicholas Wade, "Brain of Einstein Continues Peregrinations," *Science*, July 31, 1981; and interviews conducted with Thomas Harvey, Virginia Faris, Joe Collison and Cheryl Schimmel through December 2000. The quote from Raye Harvey that her husband did little with the brain while they were together came from Michael Paterniti, *Driving Mr. Albert: A Trip Across America with Einstein's Brain* (New York: Dial Press, 2000).

Information about Nathan during this period and the difficulties of the Einstein papers project came from interviews with Robert Schulmann, one of the key editors of the project, and John Walsh, "Editorial changes for Einstein papers," *Science*, April 15, 1988, p. 278. Jamie Sayen, Louise Sayen and Thomas Bucky also provided helpful insight. All the letters from the archives, including Nathan's correspondence to Harry Zimmerman, are quoted with permission from the Albert Einstein Archives, The Jewish National & University Library, The Hebrew University of Jerusalem, Israel.

Information about the truck and Israeli soldiers that stood guard over the transfer of Einstein's archival material came from the foreword written by Freeman Dyson for *The Expanded Quotable Einstein*, collected and edited by Alice Calaprice (Princeton, N.J.: Princeton University Press, 2000).

The anecdote about the little boy giving directions to the whereabouts of Einstein's brain came from Jeff Truesdell, "The Man With Einstein's Brain," *Columbia Daily Tribune*, March 1985.

Discussion of the pieces Harvey selected to send to Marian Diamond is based on the shaded areas of Harvey's maps.

The conversation Marian Diamond had with the man in the mailroom was relayed to me in my interview with Diamond in March 2000. Scheibel described his reaction to me in our interview, and the description of the origins of the Golgi stain came from Scheibel, Stanley Finger's *Origins of*

Neuroscience and Susan A. Greenfield, *The Human Brain* (New York: Basic Books, 1997).

To describe the method of Diamond's study I relied on the details set out in Marian C. Diamond, Arnold B. Scheibel, Greer M. Murphy and Thomas Harvey, "On the Brain of a Scientist: Albert Einstein," *Experimental Neurology* (April 1985), vol. 88, no. 1.

CHAPTER TEN

The anecdote about the dinner at the Hyatt Regency and information regarding the research involving the DNA of Einstein's brain was relayed to me in interviews with Charles Boyd, professor of cell biology at the Pacific Biomedical Research Center at the University of Hawaii, in March 2000 and June 2001.

Quotes from Walter Reich's commentary on Diamond's study and the study of genius brains in general originally appeared in the *New York Times*, July 28, 1985, p. 24, and was reprinted as Walter Reich, "Picking the Brains of Genius," *San Francisco Chronicle*, August 11, 1985, p. 13.

Various media accounts that covered the Diamond report offered a sense of issues the critics raised, among them: Stephen Juan, "Einstein's Brain Was Doing the Washing," *Sydney Morning Herald*, February 8, 1990, p. 12; Dan Colburn, "Inside Einstein's Brain Scientists Count His 'Glial Cells' and Wonder if They Made a Difference," *Washington Post*, March 6, 1985, p. z17. The quote from Albert Galaburda that appeared at the time was taken from Richard A. Knox, "The search for genius: Soviet brain research was part of a long quest to find physical basis of intellectual prowess," *Boston Globe*, September 10, 1991, p. 3. The anecdote about the Einstein study drawing laughter on the lecture circuit was relayed during an interview with Bruce Ransom of the Washington University School of Medicine in January 2001. Information about Scheibel and Diamond hoping to study the brain of Andrei Sakharov came from the interview with Arnold B. Scheibel.

Details regarding Einstein's concerns about the inheritability of mental illness came from Roger Highfield and Paul Carter's *The Private Lives of Albert Einstein* and Denis Brian's *Einstein: A Life*.

Harvey's quote regarding the function of glial cells came from Dan Colburn, "Inside Einstein's Brain Scientists Count His 'Glial Cells' and Wonder if They Made a Difference," *Washington Post*, March 6, 1985, p. z17. Interviews with Harvey and correspondence from the Albert Einstein Archives also supplied the details to assemble this chapter.

Part of the information to explain Nathan's dealings with reporter Gina Maranto and the answers Harvey gave at the time came from Maranto's article "The Bizarre Fate of Einstein's Brain," *Discover*, May 1985.

Interviews with Cheryl Schimmel explained how Harvey reacted to the media attention around the Diamond study, and Diamond shared her own view in my interview with her in March 2000.

The description of Nathan's apartment came from details relayed to me by Jamie Sayen and Roger Richman, of the Roger Richman Agency Inc. in Beverly Hills, California, both of whom visited Nathan there at various points. An interview in June 2001 with Jeff Truesdell, the former reporter with the *Columbia Daily Tribune* who corresponded with Nathan in 1985 and is now editor of the *Orlando Weekly*, rounded out details of their exchange.

CHAPTER ELEVEN

To reconstruct events in this chapter I relied on interviews with Charles Boyd, Robert Schulmann, Evelyn Einstein (March 2000 and January 2001), and Hilda Jost (June 2001), wife of the late physicist Res Jost, now living in Switzerland.

Roger Highfield and Paul Carter's *The Private Lives of Albert Einstein* and Denis Brian's *Einstein: A Life* provided valuable details to assemble the section dealing with Einstein's relationships with women. Robert Schulmann, Thomas Bucky and Gillett Griffin, professor of art history at Princeton University who became acquainted with Einstein a few years before the physicist's death, also supplied information on this subject. Einstein's 1900 poem to Mileva Maric was taken from the Einstein letters, which were reproduced in Jack Katznell, Associated Press, "Letters shatter Einstein's popular image, show him as unkind husband," *Fort Worth Star-Telegram*, November 17, 1996.

The quote from Einstein's letter setting out the stern conditions under which he would continue his marriage to Mileva Maric came from Dinia Smith, "Einstein letters up for auction: More than 400 letters provide an extraordinary glimpse into the scientist's marriage," *New York Times*, reprinted in the *Globe and Mail*, November 12, 1996, p. D1.

The quote from Christie's manuscripts expert was taken from "Einstein's letters sold at auction, they reveal tender and dark sides of genius," *New York Times News Service*, and reprinted in the *State Journal-Register* (Springfield, Illinois), December 1, 1996, p. 52.

Information about the plans of the California company called Stargene came from Lori Andrews and Dorothy Nelkin, *Body Bazaar: The Market for Human Tissue in the Biotechnology Age* (New York: Crown Publishers, 2001), and Roger Highfield, "For sale: the very stuff of genius," *Daily Telegraph*, January 23, 1993, p. 2.

To describe the recipe for extracting DNA from a banana I drew inspiration from "Make a banana split into DNA," *Current Science*, September 8, 2000, p. 8, from "How to Extract DNA from Anything Living," posted online by the *Genetic Science Learning Centre*, University of Utah, and from my interview with Charles Boyd.

CHAPTER TWELVE

Information in this chapter came from interviews conducted with Thomas Harvey and Louis Fishman, and in February 2000 with Ahilleas Maurellis, Harvey's former roommate, now a physicist at Amsterdam's Institute for Atomic and Molecular Physics.

Quotes and details about Harvey's concerns over the fate of the brain at this time came from Scott McCartney, "The Hidden Secrets of Einstein's Brain Are Still a Mystery—but a Kansas Doctor Remains Interested, and He Retains the Remains in His Closet," *Wall Street Journal*, May 5, 1994, p. A1.

In discussing the history and philosophy of relics and curses, I relied on various sources for details and inspiration, among them: Jonathan Freedland, "In the name of Science," *The Guardian*, December 17, 1994, p. T010; Thomas Head, "The Cult of Saints and Their Relics," posted online for *The Online Reference Book for Medieval Studies* (copyright 1999); David

Keys, "Curse of the mummy's tomb invented by Victorian writers," *The Independent*, December 31, 2000, p. 3; and Joe Nickell, "Curses: Foiled again," *The Skeptical Inquirer*, November 1, 1999, p. 16.

Some of the information about Arthur Jensen was relayed to me in my interview with Jensen in February 2001. I also relied on details from: Mary Riddell, "Is This Man Truly the World's Most Loathsome Scientist?" *Daily Mail*, September 17, 1999, p. 30; the essay on Arthur Jensen posted on the *History of Influences in the Development of Intelligence Theory and Testing*, a website directed by Jonathan Plucker, assistant professor of learning and cognition at the University of Indiana; and Arthur Jensen, *The g Factor: The Science of Mental Ability* (Westport, CT: Praeger Publishers, 1998).

Britt Anderson, neurologist at the University of Alabama at Birmingham, provided wonderful scientific background and personal details of his experience during interviews in June 1999, and in February and March 2000.

In describing Harvey's portrayal on celluloid, I drew on the work of the powerful documentary film entitled *Einstein's Brain*, directed by Kevin Hull for the BBC's *Arena Relics Series* in 1994, as well as interviews with Evelyn Einstein, Charles Boyd and Cheryl Schimmel. The detail that Kenji Sugimoto keeps his piece of the brain in a tea canister came from Merrill Goozner, "Japanese Professor Can't Get Enough Einstein: Mathematician Adds Genius' Brain to His Collection," *Chicago Tribune*, May 29, 1994, p. 11.

In July 2000, Harvey provided copies of letters sent to him by strangers inquiring about the brain, as well as permission to quote from them.

Interviews in June 2001 with William Katz, now head of Information Technology at the E and E Display Group, and Chris Cottrell, a master extruder at the same plant, provided wonderful details concerning Harvey's acts of charity and work ethic during his Lawrence, Kansas, years. Louis Fishman was also very helpful in this respect.

To summarize the work on brain cell density in men and women published by Sandra Witelson, professor of psychiatry and behavioural neuroscience at McMaster University, I relied on S. F. Witelson, I. I. Glezer and D. L. Kigar, "Women have greater density of neurons in the posterior temporal cortex," *Journal of Neuroscience* (May 1995), vol. 15, and Joseph Hall, "Why girls are better with words: McMaster scientist tests brain cells," *Toronto Star*, May 16, 1995, p. A8.

The opening statement about Anderson's report came from B. Anderson and T. Harvey, "Alterations in cortical thickness and neuronal density in the frontal cortex of Albert Einstein," *Neuroscience Letters* (June 7, 1996), vol. 210.

CHAPTER THIRTEEN

Details for the description of the reception area outside Sandra Witelson's office and Witelson's reaction to the fax Harvey sent her came from my visit and interview with Witelson in preparing the article "Decoded in Canada: Einstein's Brain," *Globe and Mail*, June 18, 1999, p. A1.

Witelson's quote about appearing on television with Gloria Steinem came from Lesley Kruger, "Brainstorm," *Chatelaine*, December 1, 1995, p. 72.

To reconstruct events and Harvey's experiences in this chapter interviews were conducted with Harvey and with Francine Benes (April 2000), director of the Harvard Brain Tissue Resource Center at McLean Hospital in Belmont, Massachusetts. Also helpful was Scott McCartney's 1994 article in the *Wall Street Journal*.

The item about Harvard University's worries about "Einsteinmania" in 1936 came from Denis Brian's *Einstein: A Life*.

In describing the 1994 report from Russian scientists about Lenin's brain, I consulted the many news articles written on the topic at the time. Quotes from Oleg Adrianov, director of the Moscow Brain Institute, came from Andrew Higgins, "Vladimir's Brain," *The Age*, November 20, 1993, p. 5, and Jay Ingram, "Why 70-year study of Lenin's brain wasn't too bright," *Toronto Star*, January 23, 1994, p. E9.

Information about the post-mortem brain of Ronnie Kray came from Derek Brown, "Head case rights and wrongs: Someone's been taking brains from the dead and the surviving Krays are not very happy about it," *The Guardian*, June 7, 1997, p. 5, and Steve Boggan, "Kray was brainless when laid to rest," *The Independent*, June 5, 1997, p. 3.

Details about the case of Jeffery Dahmer were compiled from Richard P. Jones, "Mother Wants Neurosurgeon to Investigate, Father Fights Study of Dahmer's Brain," *Milwaukee Journal Sentinel*, October 4, 1995, p. 1, and Doug Moe, "Secrets of Dahmer Brain Will Continue," *Capital Times*, June 7, 1997, p. 2A.

Information to re-create the thoughts and involvement of Gregory Stock, director of the Program on Medicine, Technology and Society and professor in the Department of Neuropsychiatry and Biobehaviour at the School of Medicine, University of California at Los Angeles, were relayed in an interview with Gregory Stock in March 2001. Background information on Stock's accomplishments came from Elizabeth Mehren, "Book of Questions: When Musings Turn into Money," *Los Angeles Times*, September 20, 1987, p. 1. Both Stock and Harvey granted permission to quote from the correspondence between them. Interviews with Robert Harvey also provided details for this section.

Einstein's views on age and solitude came from *Albert Einstein: Out of My Later Years: The Scientist, Philosopher and Man Portrayed in His Own Words*, copyright the Estate of Albert Einstein (New York: Wings Books, 1996).

Details about the Visible Human Project came from Lori Andrews and Dorothy Nelkin's *Body Bazaar: The Market for Human Tissue in the Biotechnology Age*; "Visible Human Male," in *CD Computing News*, October 1, 1997; and David Brown, "The Visible Human Project: A Slice of Life," *Washington Post*, January 13, 1999, p. H01.

News that Michael Jackson might be interested in purchasing the eyes of Albert Einstein was reported in Jonathan Freedland, "Michael Jackson sets sights on Einstein's eyes," *The Guardian*, December 17, 1994, p. 1. Henry Abrams's response to allegations that he planned to sell Einstein's eyes was relayed in an interview in February 2000.

Details about the relationship between Roger Richman, founder of the Roger Richman Agency in Beverly Hills, California, and the Einstein estate, including Richman's relationship with Otto Nathan, was relayed in an interview with Richman in March 2001.

CHAPTER FOURTEEN

Information relayed in interviews with Thomas Harvey, Ahilleas Maurellis and Robert Harvey helped in assembling the early sections of this chapter, as did visits to Harvey's home in Titusville in February and July 2000.

To explain the shortage of post-mortem brains for study I relied on interviews with Britt Anderson of the University of Alabama, Marian Diamond at Berkeley and Francine Benes of the Harvard Brain Tissue

Resource Center. Also helpful were Victor Cohn, "Wanted: Normal Brains (If Any Can Be Found)," *Washington Post*, January 24, 1989, p. 25; Jon Marcus, "Researchers Need More Brains," *Peoria Journal Star*, July 7, 1995, p. A5; Candace Gibson, "Death of the Autopsy," *Globe and Mail*, October 6, 1998, c8; and Randy Hanzlick et al., "Institutional Autopsy Rates," Autopsy Committee of the College of American Pathologists, posted online by the *Archives of Internal Medicine* (June 8, 1998), vol. 158, no. 11.

Information about Sandra Witelson's thoughts upon receiving Harvey's invitation to study Einstein's brain was relayed in an interview conducted for the *Globe and Mail* in June 1999.

In chronicling the early years of Sandra Witelson's work, Robert Pool's book *Eve's Rib: Searching for the Biological Roots of Sex Differences* (New York: Crown Publishers, 1994) provided invaluable detail and perspective. Other helpful background information came from: Lesley Kruger, "Brainstorm," *Chatelaine*, December 1, 1995, p. 72; Mark Nichols, "Boys, Girls, and Brainpower: The sexes differ in more than appearance," *Maclean's*, January 22, 1996, p. 49; Tiffany Boyd, "Picking the Brain of Albert Einstein," *The Silhouette*, McMaster University's student newspaper, January 20, 2000; and "Mapping the Mysteries of the Mind: Dr. Sandra Witelson's research makes headlines literally and figuratively," in the *McMaster Times*, Fall 1998.

Details about Witelson's former department head, Donald Hebb, came from Raymond M. Klein, "The Hebb legacy," *Canadian Journal of Experimental Psychology*, March 3, 1999.

In referring to some of Witelson's earlier research, I consulted her published papers: "Sex and the Single Hemisphere: specialization of the right hemisphere for spatial processing," *Science*, July 30, 1976; "Developmental dyslexia: two right hemispheres and none left," *Science*, January 21, 1977; "The brain connection: The corpus callosum is larger in left handers," *Science*, August 16, 1985; "Hand and sex differences in the isthmus and genu of the human corpus callosum: A postmortem morphological study," *Brain* (June 1989), vol. 112, no. 3; C. M. McCormick, S. F. Witelson, E. Kingstone, "Left-handedness in homosexual men and women: neuroendocrine implications," *Psychoneuroendocrinology* (1990), vol. 15, no. 1; S. F. Witelson and R. S. Nowakowski, "Left out axons make men right," *Neuropsychologia*, (1991), vol. 29, no. 4; C. M. McCormick and S. F.

Witelson, "A cognitive profile of homosexual men compared to heterosexual men and women," *Psychoneuroendocrinology* (1991), vol. 16, no. 6;
"Neural mosaicism: sexual differentiation of the human temporo-parietal
region for functional asymmetry," *Psychoneuroendocrinology* (1991), vol. 16,
no. 1–3; S. F. Witelson and D. L. Kigar, "Sylvian fissure morphology and
asymmetry in men and women: bilateral differences in relation to handedness in men," *Journal of Comparative Neurology* (September 1992), vol.
323, no. 3.

News articles that provided background on the public interpretations of
Witelson's work included: "Teaching approach should reflect brain differ-
ences, professor says," *Globe and Mail*, March 6, 1979, p. 3; Keay Davidson,
"Nature vs. Nurture: A wave of recent studies would have us believe that
our destinies are shaped at birth. But is there more politics than science in
this research?" *San Francisco Examiner*, January 20, 1991, p. 111; Paul Taylor,
"Age shrinks part of brain only in men," *Globe and Mail*, July 18, 1991, p.
A2; Paul Taylor, "Cognitive profile of homosexuals unique," *Globe and Mail*,
January 31, 1992, p. A5; Paul Taylor, "Brains of gay, straight men show differences: Canadian researchers report major advance in examination of cortex," *Globe and Mail*, November 17, 1994, p. A6; Paul Taylor, "Brains of
righties, lefties chemically different," *Globe and Mail*, November 18, 1994,
p. A10; and Daniel Goleman, "Right and Left Brain: Fact and Fiction," *New
York Times*, reprinted in the *San Francisco Chronicle*, October 20, 1985, p. 17.

Information on handedness, learning and premature birth was taken
from: G. Ross, E. G. Lipper and P. A. Auld, "Hand Preference of Four-Year-
Old Children: Its Relationship to Premature Birth and Neurodevelopmental
Outcome," *Developmental Medicine and Child Neurology* (October 1987), vol.
29, no. 5; G. Ross, E. G. Lipper and P. A. Auld, "Hand preference, prema-
turity and developmental outcome at school age," *Neuropsychologia* (May
1992), vol. 30, no. 5; and S. Saigal et al., "Non–right handedness among
ELBW and term children at eight in relation to cognitive function and
school performance," *Developmental Medicine and Child Neurology* (May
1992), vol. 34, no. 5.

Details about Witelson's grants from the U.S. National Institutes of
Health were supplied by the NIH archives under the file "Awards to
Sandra Witelson, Fiscal Years 1976 to 1999." Arnold Scheibel relayed

information about visiting Witelson's lab for the NIH during an interview in February 2000.

Details relating to Witelson's collection of normal brains came from an interview with Peter McCulloch, medical oncologist with the Hamilton Regional Cancer Centre, in March 2001, and from Sandra F. Witelson and Peter B. McCulloch, "Premortem and Postmortem Measurement to Study Structure with Function: A Human Brain Collection," *Schizophrenia Bulletin* (1991), vol. 17, no. 4.

Information about Witelson's phone call to Sidney Schulman of the University of Chicago was relayed during my interview with Schulman in April 2000.

The notion that even at the age of nine Einstein was not "fluent" in his speech came from Ronald Clark's 1973 biography *Einstein: The Life and Times*.

Witelson's recollection of seeing the brain for the first time in the trunk of Harvey's car came from interviews with Harvey and from "Gray Matters: You're no Einstein, says a Canadian scientist and here's why," *People*, August 30, 1999, p. 126.

Quotes from Witelson in regards to holding Einstein's brain were taken from my *Globe and Mail* interview with her in June 1999. Her comment in reference to noticing something remarkable about the parietal lobes was relayed during an interview with Thomas Harvey. Witelson's remark comparing the route of Einstein's Sylvian fissure to seeing eyebrows beneath the eyes came from Alison Motluk, "Dicing with Albert," *New Scientist*, March 18, 2000, p. 43.

The discovery of the Sylvian fissure was detailed in Stanley Finger's *Origins of Neuroscience*.

Information concerning Witelson's decision to travel to Jerusalem was relayed during the June 1999 interview and discussed with Peter McCulloch.

Details about the life and death of Henry Witelson came from Sandra Witelson, "In Memoriam Class of '63," and was posted online by *Alumni Connections*, Faculty of Medicine, McGill University.

Einstein's reaction to his father's death and the letter his father once wrote on his son's behalf came from Banesh Hoffman, with Helen Dukas, *Albert Einstein: Creator and Rebel* (New York: Viking Press, 1973).

References regarding the Witelson paper on Einstein's brain came from Sandra F. Witelson, Debra L. Kigar and Thomas Harvey, "The Exceptional Brain of Albert Einstein," *The Lancet*, June 19, 1999, vol. 353.

CHAPTER FIFTEEN

Interviews with Thomas Harvey and Evelyn Einstein and passages from Michael Paterniti, *Driving Mr. Albert: A Trip Across America with Einstein's Brain* (New York: Dial Press, 2000), provided information to describe their encounter.

Details about Harvey's visit to Marian Diamond's lab were relayed in the interview with Marian Diamond, March 2000. Latest studies referring to the power of glial cells to help neurons make connections came from Murali Krishna Temburni and Michele H. Jacob, "New Functions for Glia in the Brain," *Proceedings of the National Academy of Sciences of the United States of America*, March 27, 2001, and from Masai Iino et al., "Glia synapse interaction," and Vittorio Gallo and Ramesh Chittajally, "Unwrapping Glial Cells from the Synapse: What Lies Inside?" both of which were published in *Science*, May 4, 2001. The interview with Bruce Ransom of Washington University in January 2001 also supplied information about scientists' new regard for glial cells.

The description of events in Sandra Witelson's office the day before her paper was published in *The Lancet* was gathered during my visit there to conduct an interview for the *Globe and Mail*. The quote that Witelson's Einstein study was the biggest media story McMaster University had ever seen came from "Researcher finds differences in Albert Einstein's brain," *McMaster Courier Online*, August 16, 1999. All information about the paper itself came from Sandra F. Witelson, Debra L. Kigar and Thomas Harvey, "The Exceptional Brain of Albert Einstein," *The Lancet*, June 19, 1999, vol. 353.

To describe the optimism in neuroscience, and for quotes related to the topic, inspiration came from an interview with Donald Stuss, director of the Rotman Research Institute in Toronto, in March 2000, conducted during a conference on the frontal lobes that I attended while researching "They can (almost) read your mind," *Globe and Mail*, March 25, 2000, p. A15. As well, the quote from Israel Lederhendler, chief of behavioural neuroscience at the U.S. National Institutes of Mental Health, came from by M. Mitchell Waldrop, "Cognitive Neuroscience: a world with a future," *Science*, September 24, 1993.

Background on Rudolph Wagner's study of Karl Friedrich Gauss in 1860 came from Stanley Finger's *Origins of Neuroscience*.

Information about Witelson's presentation to the Society of Neuroscience in Miami in 1999, and discussion surrounding the recent connection between brain size and ability, was relayed during an interview with Tomas Paus, neuroscientist with the Montreal Neurological Institute, in March 2000.

Details about the studies that reflect the reemergence of brain size as a study topic were drawn from various sources, among them: Maguire et al., "Recalling routes around London: activation of the right hippocampus in taxi drivers," *Journal of Neuroscience* (September 15, 1997), vol. 17, no. 18; Maguire et al., "Navigation-related structural change in the hippocampi of taxi drivers," *Proceedings of the National Academy of Sciences* (April 11, 2000), vol. 97, no. 8; Alan Freeman, "Knowledgeable cabbies get big heads navigating London's 17,000 routes," *Globe and Mail*, March 15, 2000, p. A1; G. Schlaug et al., "Increased corpus callosum size in musicians," *Neuropsychologia* (August 1995), vol. 33, no. 8; A. L. Reiss et al., "Brain Development, Gender and IQ in Children: A Volumetric Imaging Study," *Brain* (October 1996), vol. 119, no. 5; M. D. Reynolds et al., "Small head size is related to low Mini-Mental State Examination in a community sample of nondemented older adults," *Neurology* (July 13, 1999), vol. 53, no. 1; Alasdair Palmer, "He's got a better memory than us: New research suggests that one part of an aborigine's brain is 25 per cent bigger than a European's . . ." *Sunday Telegraph*, November 19, 2000, p. 27. The number of MRI studies linking brain size to IQ came from Arthur Jensen's 1998 book *The g Factor*.

To describe the reaction to Witelson's study on Einstein's brain I also relied on information from Steven Pinker, "His Brain Measured Up," *New York Times*, June 24, 1999, p. 27, and on interviews with: Doreen Kimura, behavioural psychologist and visiting professor at Simon Fraser University in Burnaby, British Columbia, in April 2001; Francine Benes of the Harvard Brain Tissue Resource Center; Roberto Salvatori of Johns Hopkins; and

Albert Galaburda of Harvard Medical School. Salvatori, Galaburda and Jay A. Seitz, psychology professor at City University in New York, also wrote letters to *The Lancet* that appeared on November 20, 1999, vol. 354, no. 9192.

The anecdote about the University of Rochester promoting a story related to multiple sclerosis with the angle that a television character suffers from the condition came from a press release entitled "Doctors Test Memory Drug for Multiple Sclerosis Patients," June 18, 2001, issued by the public relations department of the University of Rochester School of Medicine.

The quote about the impact of celebrity science came from Frederick E. Lepore, "Dissecting Genius: Einstein's Brain and the Search for the Neural Basis of Intellect," *Cerebrum* (Winter 2001), vol. 3, no. 1.

Information about Sandra Witelson being named to the Albert Einstein–Irving Zucker Chair in Neuroscience at McMaster University was relayed during an interview for my article "Brain researcher to get chair," *Globe and Mail*, June 23, 1999, p. A7. Details about the event confirming her appointment came from Stewart Brown, "Decade of the Brain Continues at Mac," *Hamilton Spectator*, June 24, 1999, p. A12, which also included the quote from Hamilton businessman and philanthropist Irving Zucker.

Details about the popular perceptions about Witelson's involvement with Einstein's brain came from Gayle MacDonald, "Brain Power: Znaimer throws a dinner party. For $3,000, one can listen as filmmakers, techies and designers take on the future," *Globe and Mail*, January 26, 2000, p. R5; David Akin, "With your left hand, you can point to God: Owner of Einstein's brain finds where our beliefs reside," *National Post*, June 21, 2001, p. B5. Information about the details listed as part of Sandra Witelson's professional thumbnail sketch for the TEDCity conference came from "Dr. Sandra F. Witelson, Albert Einstein–Irving Zucker Chair in Neuroscience, McMaster University," on the *TEDCity 2000 Presenters* website (www.tedcity.com/speakers/witelson.html). Elliot Krauss, chief pathologist at Princeton Medical Center, provided information on his relationship with Thomas Harvey and his involvement with the brain in February 2000.

ACKNOWLEDGEMENTS

SO MANY PEOPLE gave willingly of their time and thoughts in the course of my research for this book. Faced with the queries of a stranger, most of them amazed me with the courtesy and enthusiasm of their replies. Without them, this book could not have been written. The first thank-you goes to Thomas Stoltz Harvey. A man of inordinate patience, he put up with nearly infinite and sometimes difficult questions during my visits and phone calls, yet always greeted me with the same warmth and cheer. I'm deeply grateful for all the references, letters and photographs he sent, and that he agreed to help set the historical record straight. I also owe a great debt to Barbara Wolff at the Albert Einstein Archives at the Hebrew University in Jerusalem for digging up the letters that offered an untold version of events and permission to quote from them.

Of the former Princeton Hospital doctors who shared their memories, I thank Robert Lewis, David Rose, Benjamin Wright and, in particular, Henry Abrams. Special thanks also goes to the gentleman surgeon D. Barton Stevens, for his guided tour, recollections and encouragement. The prompt and kind assistance of Barbara Ochalski, Princeton Medical Center's keeper of records supreme, simplified so many tasks. Thanks as well to former Princetonian and author Katrina R. Mason for passing on colourful details of local history.

Cheryl Schimmel, Virginia Faris, Joe Collison, William Katz, Chris Cottrell, Bobbi Fishman and especially Ahilleas Maurellis and Princeton internist Louis Fishman took the time to impart the less public dimensions of their friend Thomas Harvey. I would also like to acknowledge the insight Arthur Harvey offered and the information Frances Bermudez provided. I want to thank Robert Harvey in particular, for sharing the vantage point of his youth, the contents of a certain cardboard box and a fine couple of days on Nassau Street.

For information relating to the late Harry Zimmerman, I thank Roberto Salvatori, Robert Terry and Asao Hirano. As well, a special thanks to radiologist Bertram Lincoln Pear of Denver, Colorado, for his thoughtfulness in sharing copies of letters and articles in his own files. For information relating to Albert Einstein and Otto Nathan, I thank Thomas Bucky, Gillett Griffin, Louise Sayen, Jamie Sayen, Roger Richman and Hilda Jost. Robert Schulmann, professor of German history and one of the key editors of the Einstein Papers Project, now at the California Institute of Technology, deserves special mention for generously sharing his rich insight so that I might gain a fuller appreciation of deeper forces at work in this story. I am also truly indebted to Evelyn Einstein for sharing so much.

Many of the scientists who worked on Einstein's brain recounted their experiences with meticulous detail and candour. I thank Sidney Schulman, Marian Diamond, Britt Anderson, Jorge Columbo, Elliot Krauss, Arnold Scheibel, who also reviewed a certain portion of this manuscript, and Charles Boyd, who mailed relevant articles and offered thoughtful perspectives. Others indirectly involved with Einstein's brain also deserve my thanks, in particular Louis D. Boshes, Arthur Jensen, Gregory Stock and Peter McCulloch. I'm also grateful to Tomas Paus, Albert Galaburda, Doreen Kimura, Wally Welker, Inge Sigglekow, Bruce Ransom and Stuart Fischoff for their valuable input.

Special thanks to pathologist Charles Lee at the Coroner's Office Forensic Pathology Unit for the show-and-tell at Ontario's busiest morgue and advising me that skipping breakfast was perhaps a good idea. I am also indebted to Charles Dyer, astrophysics professor at the University of Toronto, for trying to drum a little Einsteinian theory into my innumerate mind and his valuable critique of my interpretation. To neuropsychiatrist Shitij Kapur of the University of Toronto and the Centre for Addiction and Mental Health, I deeply appreciate the time taken to discuss the big picture and review a draft version of this manuscript, though I must stress that any errors contained within the body of this work are my own.

I owe a great deal to the work of the many writers and reporters who, at various points in the meandering afterlife of Einstein's brain, produced articles that now represent an unofficial archive of the organ's history—among them Steven Levy, Jeff Truesdell, Gina Maranto and Michael Paterniti. As well, I relied on various biographies to describe certain events in this book and understand something of Einstein as both an icon of science and a human being. I give special credit to Denis Brian's *Einstein: A Life*; *The Private Lives of Albert Einstein* by Roger Highfield and Paul Carter, and Ronald Clark's *Einstein: The Life and Times.*

This book began when Jackie Kaiser, then a senior editor at Penguin Canada, called out of the blue. She'd read my story about Sandra Witelson's study of the brain in the *Globe and Mail* and believed that there was an untold story here about the workings of science. Her instinct inspired me, and for that I'm grateful.

Thanks also to my agent Dean Cooke, who managed spontaneous pep talks when I needed them, and to Catherine Marjoribanks, who tolerated the frustrating quirks of my schedule with understanding, and reviewed every line with an eagle eye. One of the great pleasures in producing this manuscript was the privilege of working with an editor whose sharp mind and keen ear shaped this into a better book; I'm terribly fortunate to have had the considerable assets of Meg Masters in my corner.

Many good friends have supported me through these last two years, offering input and generally allowing me to sound off at any given moment. I'm especially grateful to Derek Raymaker, who read an early draft. Thanks to Warren Kinsella for his advice, to my colleague Paul Taylor who first heard about the Einstein study and guessed I would be interested in it, to Erin Elder for her efforts on the photo front and to my editors at the *Globe and Mail* who allowed me the time to execute this project. Thanks as well to John Ibbitson who encouraged me to write a book in the first place, although there were moments I cursed him.

My family carried me through the ups and downs of this project with love and support. Thank you to James and Jeanette Rouse for their encouragement and the *Creator & Rebel* they found in their basement. Thank you to my brother Conrad for his sweet calls to see if I was still alive, to my brother Kevin for bringing good Chi on Mondays, and to the Clutterbuck clan—George, for those mysterious electronic flowers, Katelyn, Jared and in particular Christopher, one of my very first readers. I'm especially thankful for my superwoman sister Christine, who never stops giving. To my remarkable parents, Dudley and Thelma Abraham—no pint or gallon pot could hold my gratitude for all that you've done.

The last word of thanks is to Stephen Rouse, patient husband, best friend and meticulous editor all rolled into one. For the nights he read and reread, for his unflagging support and for walking beside me every step of the way—my debt is too steep to ever repay.

Index

Abrams, Henry, 39–40, 50,
    274–76
  AE's death and, 64–65, 66–68
Abrams, Mark, 40
acetylcholine, 80
Ackerman, Lauren, 339
Adrianov, Oleg, 261
aging, 269
Albert Einstein Archives, xiii
Albert Einstein College of
    Medicine, 7, 71, 186, 260
Albert Einstein–Irving Zucker
    Chair in Neuroscience, 330
Ali, Muhammad, 329
Alzheimer, Alois, 137
Alzheimer's disease, 77, 239,
    240, 284, 344
American Anthropometric
    Society, 79
American Hospital Association, 38
American Medical Association,
    44, 157
American Neurological
    Asssociation, 79
amyotrophic lateral sclerosis, 2,
    285, 329
Anderson, Britt, 237–41, 252–57,
    280, 325, 334

aneurysm, 48, 49, 59–60, 70,
    194, 198, 220, 335
Anti-Defamation League, 278
Archer, Nicholas, 47
Aron, Michael, 143–44, 157, 170
atomic bombs, 10, 33–34
atomic theory, 16
autopsy, 42, 44, 54–63, 72–73,
    126–27, 184, 186, 208. See
    also brain, studies of
  on AE, permission for, 203,
    246
  brain and, 285–86
  organ donation and, 293
  rates, 285

Bailey, Percival, 92, 106–8, 110,
    115, 137
Baldasaria, Dina, 44
Banting, Frederick, 320
baptism, 226
Barron, Donald, 113, 114
Benes, Francine, 261
Berger, Hans, 27
Berkeley, University of
    California, 125, 128, 162, 221
Besso, Michele, 215
Best, Charles, 320

birth control, 43, 242

body, human
  parts as relics, 65–68, 232,
    265, 270
  as property, 72, 263, 293

body snatchers, 72, 232, 286

Bonin, Gerhardt von, 90, 92,
  106, 110, 115, 133–38

Book of Questions, 268

Booth, Albie, 32

Boshes, Louis D., 107–8

Boston Globe, 197

Boyd, Charles, 193–95,
  198–200, 211, 212–15,
  227–30, 312, 335–36

brain, 60–63, 101–2, 127–39.
  See also neuropathology
  AE's, description of, 92–95,
    301–4, 306–7
  AE's, microscopic slides of,
    97, 99–106, 110–11,
    114–15, 270–73
  AE's, money for, 249, 266–72,
    279
  AE's, "ownership" of, 331–32
  AE's, permission for, xii, 18,
    54, 66–75, 203–5, 206–10,
    276, 311, 326, 346
  AE's, photographs of, 92–94,
    96, 300–302, 307, 309, 316
  AE's, promise about, 73, 75,
    83–85, 103, 108–10, 162,
    168–69, 174, 176, 183,
    202–4, 341
  AE's, reconstructing, 272–77

age and, 19, 314
Broca's area, 304, 344
cells, counting, 114, 131–33,
  190–92, 254
cells, staining, 27, 99, 189–90,
  253, 256
chemicals and, 80, 102, 132,
  142–43
collections, 100, 126–27, 198,
  241, 260, 280, 284, 326
corpus collosum, 8, 128,
  293–96, 302, 323
cortex, 12, 27–30, 80, 99,
  128, 131, 135, 137, 256
cranial capacity, 77
dendrites, 7–8, 130–31, 239
diseases, 200–201, 285
donation of, 285, 286, 292–93
drugs, 102, 128, 142–43
environment and, 314, 315
evolution of, 134–35
fetal, 3–4, 7–8, 295–96
gender and, 254, 258–59,
  288, 289–90, 292, 294, 323
glia, 4, 131–33, 138, 187,
  191, 196, 197, 315
hemispheres, 289–90
hormones and, 295–96
intelligence and (*See* genius
  and intelligence)
left/right handedness and,
  294–98
lobes, 135–36, 255–56, 302
mapping, 137, 303
memory and, 80

of mice and rats, 129–31, 238
neuron/glial-cell ratio, 325
neurons, 7–8, 27, 130–32,
    142, 165, 196, 239,
    288–89, 314
scans, 141, 285
sex and (*See* brain, gender and)
sexual orientation and, 288,
    294–95
shortage, 284–86
size, 130, 322–25
split-, 142
studies of, 26–30, 34–35,
    77–81, 88–89, 91–95,
    101–2, 203, 320–25 (*See
    also* Diamond, Marian; Vogt,
    Oskar; Witelson, Sandra)
Sylvian fissure, 303–4, 307,
    325, 327, 328
synapse, 190
thalamus, 105, 111, 114–15
Brian, Denis, 49, 172, 218
Broca, Paul, 28, 80, 101, 190
Brodmann, Korbinian, 94–95,
    137, 190
Brooklyn Jewish Hospital, 48
Brower, Anne Bonine, 93
Brown, Lesley, 320
Brown, Robert, 15
Brownian motion, 15
Buck, Pearl, 36
Bucky, Gustav, 6, 51
Bucky, Thomas, 6, 20, 49–50,
    51, 74, 217
Bush, George, 320

Cade, John, 102
California Institute of
    Technology, 163, 172, 179
cancer and longevity, 291–92
Carnarvon, Lord, 232
Carson, Johnny, 162
Carter, Paul, 52, 73, 147, 148,
    201, 210, 217
celebrity, 278–79, 332–33
    AE's, 5–7, 68, 70, 126, 158–59
Celebrity Rights Act, California,
    278
celloidin, 98, 99, 103, 190, 253,
    273
Cellular Pathology, 59
Census Bureau, 231
Chaplin, Charlie, 163, 333
Chicago Daily Tribune, 83
chlorpromazine, 102
Christie's auction house, 224
Churchill, Winston, 179, 297
City Lights, 163, 333
Citytv, 331
Clark, Ronald, 143, 148
Clemente, Carmen, 192, 196
Cody, Buffalo Bill, 167
cognitive psychobiology, 288
Cold War, 89, 198
collagen, 194
*Collected Papers of Albert Einstein,
    The,* 223
Collier's, 109
Collison, Joe, 168
Columbia Daily Tribune, 168, 207
Communism, 86, 89

compass, magnetic, 12
computers, 79, 141, 193, 265–66, 272, 317
Coover, Chris, 224
Cortizano, Benjamin, 47, 69
cosmological constant, 163–64
cosmology, 18
Cottrell, Chris, 251
cremation, 50, 61, 65, 72, 208
Crick, Francis, 79, 122
CT scans, 141, 285
Cushing, Harvey, 91

Dahmer, Jeffrey, 262–63
Dahmer, Lionel, 263
Darwin, Charles, 134
David Lavin Agency, 330
Dean, Guy, 40, 44, 48–49
    AE's death and, 53, 54, 67
"Decade of the Brain," 321
Delaware Poetry Society, 119
Descartes, René, 62, 79
Deutsches Hygiene-Museum, 344
Diamond, Marian, 125–39, 162, 165, 182, 234, 314–16
    AE's brain and, 186–92, 195–98, 200, 202, 204, 309, 310
"diener," 126–27
Dillinger, John, 108
Discover, 203, 204, 206, 207
discrimination. See genius and intelligence, race and
DNA, 79–80, 128, 193, 194, 199, 214, 227–30, 335

Down's syndrome, 324
Driving Mr. Albert, 310
Dukas, Helen, 119, 179, 201
    AE's death and, 65, 74
    AE's estate and, 74, 116, 117, 147, 151, 169, 176, 180, 223
    as AE's guardian, 39, 45, 47–58
    Evelyn Einstein and, 227
dyslexia, 289–90

E and E Display Group, 231, 250–51
Economist, The, 335
Edgewood Arsenal, 33
Edwards, Robert, 320
Ehrich, Wilhelm, 103
Einstein, Albert (AE)
    affairs with women, 216–19, 223
    autopsy of, 54–63
    birth of, 8, 11
    brain of (See brain; genius and intelligence)
    childhood of, 11–13, 297
    commercializing, 278–79
    death of, 20–21, 53
    early career, 305–6
    estate, xii, 74, 82, 117, 119, 147–48, 151, 176–77, 207, 219, 222–24
    Evelyn Einstein and, 216, 218–20, 224–27
    eyes, 39, 66, 67, 274–75, 335
    final days, 48–53

language learning, 12, 171,
297, 304, 345
Nobel Prize, 14, 15, 199
poetry, 218, 223
as refugee, 179
statue of, 267
theories of, 9–10, 13, 15–18,
19, 49, 163–64, 218
violin and, 12, 36, 45, 162,
179, 296
will of, 50, 56, 65, 67, 74
Einstein, Aude, 219
Einstein, Bernard, 199, 201,
214, 219
Einstein: Creator and Rebel, 306
Einstein, Eduard, 40, 52, 74,
117, 179, 199, 200–1, 226
Einstein, Elizabeth Roboz. See
Roboz, Elizabeth
Einstein, Elsa, 52, 163, 177, 179,
217, 245
Einstein, Evelyn, 199, 201, 214,
216–22, 224–27, 246, 333
Thomas Harvey and, 309,
310, 312–13
Einstein, Frieda. See Knecht,
Frieda
Einstein, Hans Albert, 40,
51–54, 156, 199, 201
AE's brain and, 72–75, 84,
109, 183
AE's letters and, 117, 119,
148, 219
AE's will and, 74
Einstein, Hermann, 4, 199, 305–6

Einstein, Klaus, 225
Einstein, Maja, 13, 40, 199
Einstein, Margot, 52, 65, 74, 147,
179
Einstein, Mileva, 52, 201, 215,
216, 219, 223
Einstein, Pauline, 3, 8, 199, 201
Einstein, Thomas, 214
Einstein in America, 74, 154
Einstein on Peace, 147
electoencephalogram, 10
electronmicroscopy, 71
Elephant Man, 275
embalming, 63
$e=mc^2$, 10, 17, 279
endorphins, 143
energy, theory of, 17
ENIAC, 79
Enriching Heredity: The Impact of the
Environment on the Brain, 315
epilepsy, 80, 102, 141–42, 289
Euclid, 13
Ewing Crematorium, 70, 72
Experimental Neurology, 192, 195,
196, 318

Fanta, Joanna, 218
Faris, Virginia, 167–68
Farquhar, John, 91
Faulkner, William, 39
Federal Bureau of Investigation,
88, 108
Federal Licensing Examination,
211–12
FedEx, 241, 252, 280

Fields, W.C., 279
Finger, Stanley, 78
Fischoff, Stuart, 332–33
Fishman, Louis, 41–42, 117–18,
    145, 212, 250, 282, 283, 312
Fitzgerald, F. Scott, 36
Flexner, Abraham, 172–73
Flint, Joyce, 263
Fox, George, 23, 25
Fox, Michael J., 329
France, Anatole, 322
Free Speech Movement, 129
Freedland, Jonathan, 232, 245
Freedman, Hyman, 286–87
Freud, Sigmund, 30

g Factor, The, 235
Gage, Phineas, 255
Galaburda, Albert, 197, 326–27
Galileo, 66, 144
Gall, Franz Joseph, 27, 63
Gallup, George, 36
Gauss, Karl Friedrich, 78, 130,
    322
Gaylord Sanatorium, 31–32, 244
Gazzaniga, Michael, 142
Gehrig, Lou, 329
genetics, 193–94, 269, 284–85,
    335. See also DNA
genius and intelligence, 28–30,
    195–98, 322–29
    AE's brain and, 9, 11, 14,
        86–87, 88, 165, 192, 236,
        306–7 (See also under brain)

brain and, 195–96, 323–24,
    327, 330, 344–45
genes and, 335
race and, 129, 234–37, 324
research, 129, 195–96,
    234–41 (See also brain,
    studies of)
Glenn, Frank, 49
Glezer, Ilya, 345
glia. See under brain
Glia, 197
Globe and Mail, xi, 318, 331
Goldman, Ron, 333
Goldstein, Kurt, 120–21
Golgi, Camillo, 189–90
Good Morning America, 174
gravity, 17
Griffin, Gillett, 218
Gynecology & Obstetrics, 198

Hamilton, Ontario, 298
handedness, 294–98
Hardy, G.H., 19
Harper, Clive, 324
Harper's, 143, 157, 175, 310, 311
Harvard Brain Tissue Resource
    Center, 260
Harvard Educational Review, 234
Harvard University, 260
Harvey, Arthur, 22, 34, 69, 104,
    282, 346
Harvey, Elizabeth, 146, 153, 282
Harvey, Elouise, 32–35, 44, 76,
    93, 282

AE's brain and, 104
marital troubles and, 118–19,
    121, 124
Harvey, Frances, 146, 282, 311
Harvey, Frances Stoltz, 23, 25,
    31, 76
Harvey, Jean, 23
Harvey, Judge Lawson, 24
Harvey, Lisa, 145, 149, 151–57
Harvey, Rachelle (Raye),
    166–67, 173, 187, 212, 231
Harvey, Robert, 34, 42, 104,
    118, 145, 282
    Gregory Stock and, 264,
        266–69
Harvey, Thomas P. (father of
    Thomas Stoltz), 22–26
Harvey, Thomas (son of Thomas
    Stoltz), 34, 211, 282
Harvey, Thomas Stoltz, 22–27,
    30–35, 140–41, 145–54,
    156–62, 211–13
    AE's brain and, xi–xiii, 70–76,
        82–115, 165, 187, 192,
        199–210, 328, 336–42
    AE's brain to Canada, xi,
        283–84, 298
    autopsy on AE and, 54–63, 66,
        67, 68, 70
    Britt Anderson and, 241, 257
    career after medicine, 231–33,
        259–63
    Cleora Wheatley and, 145,
        282–83, 308

divorce and, 121–24
    Evelyn Einstein and, 309, 310,
        312–13
    Gaylord Sanatorium and,
        31–32
    as general practitioner,
        166–69, 173–75, 211
    Gregory Stock and, 264–80
    Marian Diamond and, 314–15
    Maurellis, Ahilleas and,
        242–45, 249–50
    medical licence and, 211–12,
        231
    Michael Paterniti and, 308–14
    Princeton Hospital and, 35,
        41–44, 121–24, 311
    research report and, 112–20,
        160, 170–71, 175, 177–78,
        180–87
    Sandra Witelson and, 281,
        284, 286, 298–302, 307,
        316–17
    strangers' letters to, 247–49
    war service and, 33–34
Haymaker, Webb, 88–90, 103
heart disease, 199. See also
    aneurysm
Hebb, Donald, 288–89, 304, 328
Hebrew University, Jerusalem,
    74, 177, 180, 207, 222, 261,
    278, 279, 304
Hiden, Conrad, 39
High Alpine Daughters Institute,
    220, 224

Highfield, Roger, 52, 73, 147, 148, 201, 210, 217
Hitler, Adolf, 6, 179
Hoffman, Banesh, 306
Holton, Gerald, 147
Honda, Harriet, 44
Hope Diamond, 232
House Committee on Un-American Activities, 86, 220
Hubble, Edwin, 164
Hull, Kevin, 246
Human Genome Project, 193
Humanité, L', 67
Huntington, Robert, 263
Huntington's disease, 285

Ide, Charles, 273
imipramine, 102
Individual Liberties Association, 43, 86
intelligence. See genius and intelligence
Intelligence, 237
Iron Curtain, 89
Isocortex of Man, The, 91, 92, 94, 106, 115, 134, 151, 314, 328

Jackson, Michael, 275
Jensen, Arthur, 128–29, 234, 240
Jernigan, Joseph Paul, 266
Jost, Hilda, 215–17, 219
Jost, Res, 215–16, 218, 219
Journal of Clinical Pathology, 244
Journal of Comparative Neurology, 113

Journal of Neuroscience, 258
Journal of the American Medical Association, 113, 244

Kaelin, Kato, 333
Kahler, Alice, 218
Kaiser Wilhelm Brain Institute, 28
Kansas City Times, 174, 176, 178, 203, 204
Kant, Immanuel, 62
Katz, William, 250–52
Katzenellenbogen, Estella, 217
Kauffman, Jack, 35, 37–39, 41, 44, 53, 337
    AE's brain and, 64, 66, 68, 81
    AE's death and, 63
    Thomas Harvey's Princeton contract and, 121–23
Keller, Marta, 97–99, 102
Kigar, Debra, 299, 300, 303, 307, 345
Kimura, Doreen, 328
Kissane, John, 339
Knecht, Frieda, 116–17, 119, 201, 216, 219, 226
Krauss, Elliott, 336–41, 343–45, 347
Krauss, Mary, 339, 340
Kray, Kate, 262
Kray, Reggie, 262
Kray, Ronnie, 262
Kuhlenbeck, Hartwig, 90, 91, 103, 181, 185

Lambert, George B., 38
Lancet, The, 196, 307, 316, 317, 319, 326, 328, 344
Lawrence, Kansas, 231–33
League of Women Voters, 43
Leavenworth Prison, 173, 187
Lebach, Margerette, 217
Lederhendler, Israel, 321
"left," 296
Lenin, Vladimir Ilyich, 29–30, 71, 77, 89, 138, 198, 261
LePore, Frederick E., 329
Levy, Steven, 157–58, 159–62, 170, 310
Lewis, Robert, 122, 123
Lewy, Frederick, 34, 70, 87, 97, 103
Lewy Bodies, 35
Lieserl (AE's daughter), 223, 227
Life, 109
light, theory of, 15, 17, 172
longevity and cancer, 292–93
Lou Gerig's disease, 285, 329

"Man of the Millennium," 331
Mandela, Nelson, 331
Manual of Pathological Anatomy, A, 58
Maranto, Gina, 203, 206, 310
Maric, Mileva, 52, 117, 201
Marlboro Psychiatric Hospital, 146
Mary Magdalene, 65
Mason, Katrina, 22

Massachusetts General Hospital, 10
Mathematician's Apology, A, 19
mathematics and age, 19
Mather Funeral Home, 63
Matzke, Howard, 139, 167, 181, 182, 185, 187
Maurellis, Ahilleas, 242–45, 249–50, 281, 346
Mayer, Walther, 172, 179
McCarthyism, 51, 86, 88, 220
McCulloch, Peter, 291–93, 298, 305
McGill University, 288, 289, 305
McGraw, Hack, 38
McLean Hospital, 260, 261, 286
McMaster University, xi, 241, 257, 289, 320, 330, 331
    Medical Centre, 299
    Psychology Department, 258
media, 45, 47, 53, 158–59, 174–75
    AE's brain and, 69, 75, 76, 81, 82, 108–10, 162, 186, 195, 201–5
    AE's death and, 63, 68, 69
    Cleora Wheatley and, 282–83
    Gregory Stock and, 275
    Otto Natan and, 176, 177
    Sandra Witelson and, 258–59, 295, 318–21, 330
Medical Research Council of Canada, 300
mental disease, 80, 102, 200
meprobamate, 102

Metaman: The Merging of Humans and Machines into a Global Superorganism, 268
microtome, 94, 99
Milikan, Robert, 172
Montcalm, Louis-Joseph de, 66
Montefiore Medical Center, 1, 5, 186
    AE's brain and, 71, 77, 81, 82, 183
Montreal Neurological Institute, 88, 101, 323
Mooney, Tom, 172
Moonies, 220
Moore, Dr., 157
Moscow Brain Institute, 197, 261
Mount Wilson Observatory, 164
Mullis, Kary, 228
multiple sclerosis, 77, 329
music, 12, 200, 201
Mussolini, Benito, 88
Mutual Autopsy Society, 79

Napoleon, 65, 144
Nash, John, 151, 152
Nassau Hall, 36
Nathan, Otto, 50–51, 52, 65, 278
    AE's brain and, 74–75, 83–87, 97, 99–100, 312
    as AE's executor, 56, 108–9, 144–45, 169–71, 176–78
    AE's family and, 74, 110, 117, 119–21

    autopsy of AE and, 56, 60, 68, 345–46
    Evelyn Einstein and, 221
    Thomas Harvey's research and, 108–17, 147–52, 154–56, 180–87, 203–11
National Academy of Sciences, 267
National Institutes of Health, U.S., 290–91, 300
National Institute of Mental Health, U.S., 321
National Post, 332
Natural Born Killers, 262
nature *vs.* nurture, 12, 30, 128
Nazis, 34, 50, 66, 101, 120, 179, 305
neurodegeneration, 34–35
neuron. *See under* brain
neuropathology, 26–30, 88
Neurophysiology, 91
Neuroscience Letters, 256, 257, 318
New England Journal of Medicine, The, 32, 244, 286
New Jersey Monthly, 157, 162, 233
New Jersey Psychiatric Institute, 145
New Jersey World-Telegram and Sun, 75
New York Times, 69, 70, 72, 75, 195, 325
Newton, Isaac, 9, 10, 297
Newton's laws, 14, 16, 17
Nissen, Rudolph, 48, 57

Nissl, Franz, 27, 99
nitrocellulose, 98
Nuata, Walle, 90, 91, 110
nuclear arms, 34
nuclear fission, 37
nucleotides, 229
Nuremberg trials, 101, 122

O'Neill, Eugene, 32
Organization of Behaviour, The, 288
Orlando Weekly, 210
Ottumwa Daily Courier, 76

Pais, Abraham, 10, 50
Paramount Pictures, 310, 346
parietal operculum, 303–4
Parkinson's disease, 285, 329
Paterniti, Michael, 308–14, 346
pathology, 33, 34, 41–42,
    58–59. See also autopsy
    clinical, 35
    Princeton Hospital and, 38,
        41–42, 338–40
    warfare and, 33
Paus, Tomas, 323, 324
Pavlov, Ivan, 30
Pearl Harbor, 33
Penfield, Wilder, 80, 88, 101,
    288, 321
penicillin, 37
Penn, William, 24
People, 330–31, 333
Phil Donahue Show, 204
Philadelphia General Hospital, 35

photography of AE's brain. See
    under brain
phrenology, 27–28, 77, 80, 324,
    344
Pinker, Steven, 325
pituitary gland, 142
Planck, Max, 15
Planned Parenthood, 43
Plesch, Janos, 216
Plesch, Peter, 217
pneumo-encephalograms, 101
polio vaccine, 79
Pool, Robert, 292
premature births, 297
Presley, Elvis, 278
Princeton, New Jersey, 36–37
Princeton Hospital, xii, 35,
    37–44, 66, 76, 343
    AE in, 49–64
    AE's brain and, xi, 81, 82,
        92–95, 100, 183, 203,
        340–44
    pathology, 38, 41–42, 338–40
    Thomas Harvey's contract
        with, 121–24, 157
Princeton Packet, 37, 47, 343
Princeton Plasma Physics
    Laboratory, 37
Princeton University, 36–37, 70
    Institute for Advanced Studies,
        82, 147, 173, 180, 215
    Woodrow Wilson School of
        International Studies, 37,
        267

Princeton University Press, 151,
    154, 169–70, 177, 207
Private Lives of Albert Einstein, The,
    52, 201
psychopharmaceuticals, 102
Purvis, Melvin, 108

Quakers, 23–25, 43, 73, 174,
    242–43, 250, 283, 347
quantum theory, 15

racism. *See* genius and intelli-
    gence, race and; Nazis
Raman y Cajal, Santiago, 189–90
Ransom, Bruce, 133, 197, 315
Reeve, Christopher, 329
Reich, Walter, 195–96
relativity, theory of, 9, 17, 163
relics, body parts as, 65–68, 232,
    265, 270
Religious Society of Friends. *See*
    Quakers
Richman, Paul, 278
Richman, Roger, 278
Rieber, Winifred, 39
Robert Wood Johnson Medical
    Center, 339
Robeson, Paul, 36
Roboz, Elizabeth, 220, 222
Roger Richman Agency, 278
Rokitansky, Karl von, 58
Rolph, James, 172
Rosai, Juan, 339
Rose, Jerzy, 90

Ross, Ginny, 212
Ross, Rachelle. *See* Harvey,
    Rachelle (Raye)
Rotary Club, 43, 86, 167, 168
Rothberg, Harvey, 123–24
Rotman Research Institute, 321
Rumpel, George, 168
Rutgers University, 193, 194,
    198, 214

Saint Elizabeth, 65
Sakharov, Andrei, 197–98
Salvatori, Roberto, 2, 3, 19,
    325–26
San Francisco Chronicle, 197
Sayen, Jamie, 51, 70, 74, 75,
    154–55, 170, 183, 210
Sayen, Louise, 69, 87, 104
Scheibel, Arnold, 137–38, 165,
    189, 191–92, 197–98, 202,
    204
    Sandra Witelson and, 291
Schimmel, Cheryl, 173–75, 204,
    212, 247, 312
schizophrenia, 102, 285
Schroeder, Howard, 96
Schulman, Sidney, 100–1, 157
    microscopic slides of AE's
        brain and, 104–6, 111,
        114–15, 301
Schulmann, Robert, 87, 117,
    148, 149, 150, 156, 209
    AE's biography and, 227,
        297

AE's papers and, 169–70, 177, 199, 214–16

Evelyn Einstein and, 214–16, 221–24, 312

Science, 125, 162, 175, 244, 290, 294, 318

Scott-Brannigan, Lisa. *See* Harvey, Lisa

Seelig, Carl, 203

Seitz, Jay, 328

Shawkey, Elouise. *See* Harvey, Elouise

Sheen, Martin, 329

signature hunters, 68

Siljestrom, P.A., 322

Simpson, Nicole, 333

Simpson, O.J., 333

Sinclair, Upton, 36

Skinner, B.F., 30

Smithsonian Institution, 260

Society of Neuroscience, 141, 322, 344

space/time, 9, 10, 16–17

Spearman, Charles, 234

Sperry, Roger, 142

Spitzka, Edward Anthony, 79

St. Einstein Brigade, 148

Stachel, John, 169, 177, 222

Stalin, Joseph, 29, 89

Stargene, 228

Steptoe, Patric, 320

Stevens, D. Barton, 53

Stewart, Jimmy, 36

Stock, Gregory, 264–80

Stock, Saint Simon, 65

Stoltz, Frances. *See* Harvey, Frances Stoltz

Stoltz, Oscar, 76

Stone, Raymond, 41, 100–1, 104–5

streptomycin, 79

Stuss, Donald, 321

Sugimoto, Kenjii, 171, 246

Sylvian fissure. *See under* brain

Sylvius, François, 303

Szechenyl, Chris, 174

Tale of Two Cities, A, 54

Taylor, Paul, 318

TEDCity conference, 331

television, 37, 42, 110

Terenius, Lars, 143

Terry, Robert, 71, 77

Time, 331

Titusville, New Jersey, 283

Today Show, The, 122, 174

transplants, kidney, 79

Travelers Insurance Company, 22, 24, 26

Truesdell, Jeff, 207–10

tuberculosis, 31, 79

Tutankhamen, 29, 232

unified field theory, 19, 49, 163–64, 218

University College, London, 323

University of Alabama, 237, 256

University of California, 125, 128, 162, 221

University of Illinois, 106

University of Kansas, 242

University of Maryland, 110

University of Pennsylvania, 34, 79, 97, 102, 103, 260

University of Wisconsin, 262

U.S. Armed Forces Institute of Pathology, 88

U.S. Army, 90, 96

U.S. Chemical Warfare Service, 33

U.S. Postal Service, 111, 252

Virchow, Rudolph, 58–59, 131

Visible Human Project, 265, 272

Vogt, Cecile, 27, 28–29
  cortex research and, 137–38

Vogt, Oskar, 27, 28–30, 34, 70–71, 77, 89, 165
  cortex research and, 137–38, 190, 197, 198, 261

Wade, Nicholas, 175

Wagner, Rudolph, 78, 322

Wall Street Journal, 233–34, 240, 259, 264

War of Independence, 36

Warhol, Andy, 340

Warsaw Ghetto, 305

Washington, George, 36

Watson, James, 79

Wayne, John, 278, 279

Weldeyer, Wilhelm von, 27

Welker, Wally, 273

West Wing, The, 329

Weston, Missouri, 166–67, 173

Wheatley, Cleora, 145, 282–83, 308, 310, 346

Whitman, Walt, 260, 265

Wilson, Woodrow, 42

Winternitz, Milton, 33

Wistar Institute, 260

Witchita, Kansas, 153

Witchita Eagle and Beacon, 171

Witelson, Henry, 284, 287, 298, 305, 307

Witelson, Sandra, xi, xiii, 241, 254, 257, 258–59, 280–307
  Elliott Krauss and, 344–45
  Lancet article and, 316–17, 318–32, 334

woolly mammoth, 335

Yeshiva University, 2, 5

Zangger, Gina, 215–16, 218

Zangger, Heinrich, 215

Zimmerman, Harry, 1–3, 5–7, 9, 107, 203, 280, 325–26
  AE's brain and, 11, 18–19, 70–71, 76, 77, 81–85, 100, 183–87, 192, 260
  microscopic slides of AE's brain and, 13–14, 100, 110
  neuropathology and, 26, 27, 29, 30

Zimmerman, Miriam, 3

Znaimer, Moses, 331

Zucker, Irving, 330